Clyde Dyke 2.25

CB for Christians

Beau Colle
with Jo Colle

Foreword by Landrum P. Leavell

BROADMAN PRESS
Nashville, Tennessee

4255-76
ISBN: 0-8054-5576-0

Dewey Decimal Classification: 248.5
Subject Headings: EVANGELISTIC WORK // WITNESSING
Library of Congress Catalog Card Number: 76-45554
Printed in the United States of America

Footnotes in this book are not consecutive, but rather refer to sources in the bibliography.

Photos: John Green, Beau Colle, and the Foreign Mission Board, SBC
Illustrations: Jo Colle

Dedicated to the many Christians
who are using CB as a witnessing tool

ACKNOWLEDGMENTS

I wish to extend my heartfelt thanks to all who have contributed to the writing and preparation of this book, particularly the CB'ers around the country who have taken time to share their testimonies and experiences with Citizens Band radio, and to Mrs. Larry Sterne, my secretary, who spent many hours in typing and working with the manuscript.

BEAU COLLE
Alexandria, Louisiana

FOREWORD

I have known Beau Colle since he was a high school football star at Pascagoula, Mississippi. In fact, he led the Pascagoula team for two consecutive years as quarterback in defeating the Gulfport High School team, where I was serving as pastor! He impressed me then, and also during his outstanding football career as defensive safety for the L.S.U. Tigers.

It was a real joy to be associated with Beau on the staff of First Baptist Church, Wichita Falls, Texas. He served first as our minister of recreation, and later as minister to youth. Again, he impressed me, but this time with his ability to establish rapport with youth and adults, and his effectiveness in ministry in these areas.

Now he has chosen to write a book on a subject of tremendous interest. I believe it is a subject that needs treatment, and I appreciate the manner in which he has dealt with it. On a quail hunting trip to Texas last fall, Beau and I had the opportunity of "modulating" with all of the "good buddies" between Alexandria and Wichita Falls! I am confident that this timely book, written by the "Louisiana Canine," will be a challenge and inspiration to Christian CB operators.

LANDRUM P. LEAVELL
President, New Orleans Seminary

CONTENTS

INTRODUCTION

Confessions of a Reluctant Writer

"Lord, are you sure we should write this book?"

Yesterday–seventeen days before the deadline for this manuscript to be turned in–I sat at my kitchen table with stacks of reference books and notes in front of me and pondered that question time and again. It was not the first time the question had entered my mind, in fact, the query had visited my thoughts *frequently* from the very day Beau was asked to write a book on the Christian's use of the Citizens Band radio.

I remember that day quite well. Beau was lying in bed, scarcely able to turn from side to side because of a back injury. Two of our children and I were gathered around Beau. We were rubbing sore muscles, adjusting pillows and applying heat in one place and ice in another in an attempt to make him more comfortable. Interspersed between all the activity we were openly contemplating our summer schedule and trying to determine how Beau would fulfill each of his obligations if his back continued to plague him. Five conferences in different parts of the country were already scheduled, and it was only June!

When the phone rang and the caller asked for Beau, I silently mouthed, "Do you feel like talking?

I think it's long distance." (Neither Beau nor I can stand the mystery that surrounds an unnamed caller, particularly if he happens to be phoning long distance, so Beau did as I guessed he would–he reached for the phone.)

I could tell as he spoke he was trying hard to sound his usual jovial self, but I recognized his exhaustion after the conversation was completed. Only my realization of his pain kept me from pulling every word of the call from him.

When he caught his breath enough to tell me he had just been asked to write a book, my first reaction was to laugh–a side-splitting, uproarious laugh–Beau write a book! Incredible! I could just imagine the look on his high school English teacher's face–utter disbelief. She could scarcely get him to write a three-page book report. Even more incredible–they wanted him to write the book in less than three months. It all seemed so ridiculous. Surely he wouldn't take the matter seriously. Surely he wouldn't–*but he did.*

"Beau, are you out of your mind! How are you going to write a book between now and September? If your back improves, we are supposed to leave for Washington in three days. When we get back you'll only be home two days before you have to leave for New Mexico. What about the conference in Mississippi? What about Ridgecrest? There is still the possibility you'll have to have surgery and three months' convalescence will be required after that. Do you think this is wise?"

If I had only taken time to remember the last eleven years, I would have understood his reasoning.

Beau and I have always been exactly opposite in the way we do things. I insist on the methodical, systematic, organized approach to everything. I have to know where I am going—plan ahead; leave no detail to chance. Beau, on the other hand, is content to leave the details and the timing to the Lord and simply be ready to move whenever He gives the go ahead.

When I entered college, my plans for the future were made. I was going to study speech pathology, marry Beau, and settle down to a life of part-time career woman, full-time wife, and mother.

When Beau entered college, I thought his plans for the future were anything but definite. After three years of study, he still had no special field. Considering that I planned to marry such a muffin-head I would argue, "Beau, you have to learn how to do *something.* How are you ever going to get a job if you don't know how to do something?" Beau would reply in his ever patient way, "I don't know now, but I will when the Lord lets me know."

I couldn't understand his way of thinking. There he was, nearing the point when he would step into a dog-eat-dog world and he didn't even have the sense to fortify himself with "puppy chow." Frankly, I was bewildered and a little disgusted, but since I thought he was the "coolest" thing to ever step out of the South, I resolved that whatever he decided to do, I was going to be a part of it.

Graduation came and I found him enrolling in New Orleans Baptist Seminary. "Oh, good, Beau. You've decided on something. You're going to be a preacher."

"No, Jo. I don't know yet what God wants me to do."

"What do you mean you don't know what he wants you to do. You're going to the seminary, aren't you?"

"I only know God wants to use me. I don't know how yet. But I'll be ready whenever he is."

So this is the way it has been for eleven years: Beau has gone from the seminary to work first as activities director for First Baptist Church, Wichita Falls, Texas, and then as youth director at that same church to educational minister in Greenwood, Mississippi, young adult consultant for the state of Louisiana and now Sunday School Program Director for Louisiana. God has changed *how* he used Beau a number of times in a short span of years. He continues to work out the details; Beau continues to follow—sometimes dragging me along behind with heels dug in.

I don't know exactly how it happened. All I remember is that, once his surgery was over and I saw that he still intended to write the book, I began hearing statements like, "I'm so glad you understand, Jo. I've always believed that, if God wanted the book to be written, it would get done—with or without your help."

"My help! Who said anything about me?" Sure I like to dig things out of the library, but–put them together? That is an entirely different matter. When would I find time? What about the children? My practice? The research project for the institute!

I couldn't help being myself–reluctant and pessimistic.

"I'm not the person for a job like that! I'm not a good speaker. I never have been, and I'm not now, even after you have spoken to me, for I have a speech impediment." Remember Moses? (Ex. 3:11, 4:10) He had a problem with reluctance, too. As a matter of fact, so did Jeremiah. Both men were wary of their abilities to carry out the tasks given them by the Lord. But the Lord's command to each was clear. In essence, it was, "You will go wherever I command you and speak whatever I tell you to. And don't be afraid of the people, for I, the Lord, will be with you and see you through." (See Jer. 1:7-8)

Thank you, Lord, for supplying the strength and words we need and–please forgive one more reluctant messenger.

DISCOVERY

I had just gotten my CB. In fact, it was my first day to use it. I had saved and scrimped for months because I just "had to get a CB radio for my car."

It all started, and I got the "bug" when I rode with a friend to a meeting in south Louisiana. He was an avid "CB'er," had been for years, and still loved it. I asked him question after question, and my interest and delight continued to grow. Actually, it didn't take too much selling to get me sold, but when I was, it was "hook, line, and sinker."

The jargon of the truckers and the amusing conversation kept my interest mile after mile. It was a different world for me and brought back memories of Broderick Crawford on "Highway Patrol." Back then I thought how "neat" it would be to one day have a radio in my car. The day had finally arrived.

Traveling down Highway 71, I had quickly picked up some of the conversations going on between 18-wheelers and 4-wheelers. I learned what a "front door," "back door," and "rocking chair" were because I had lucked up and gotten in the middle of a convoy. "High Pockets" led the way (front door) and "Little John" was keeping the "back door" closed.

I had not thought about a handle for myself too much, because I was just going to listen for a while to learn. That didn't last long because when I got in the convoy, they wanted to know my "handle." The only one I had really thought about was "Louisiana Canine." My last name Colle (pronounced Collie) will give you a clue about the origin. It stuck!! "Louisiana Canine" was in the "rocking chair" with "High Pockets" and "Little John." In fact, to this day, I haven't heard anyone else that uses the handle "Canine," and I have modulated all over the South and East.

When the ice was broken, the water really flowed. I wanted to talk to anyone, anytime, and under any conditions. Someone breaks for a 10-36, and I'm the first to give it. Someone wants to find out directions to City Hall–I give them. I was, in the first few weeks of my CB life, what I later learned was called a "Channel Hawg"–a person who tries to monopolize the conversation. I didn't know any better at the time–I was just having a ball!

As the convoy continued to move down Louisiana 71 and then on to US 190, "High Pockets" went 10-7 along with "Little John." I bid them a good day and a better tomorrow and continued eastbound. Not long after this, there was a breaker for the one "Louisiana Canine." I responded to it and it was a 4-wheeler by the name of "Laredo." He too was traveling East and broke in to tell me about some ground fog conditions he had just encountered.

Laredo, I surmised, was on my front door. We exchanged mile markers and he was running about three miles ahead of me. We continued to modulate on a number of subjects from the weather to what kind of units we were using. It seemed as though he wanted to talk and, of course, so did I.

I asked where he was headed. "Superdome City" was the response. "Where's your home-20?" I continued. "Sport City" (Shreveport) came through wall to wall. The conversation continued to flow, and we exchanged work-20's. He was in the oil field business in north Louisiana, and I told him I worked with the Louisiana Baptist Convention. This immediately caught his curiosity, and he pressed for more information. I explained that I worked in the Sunday School department and was an ordained Baptist minister.

He volunteered that he and his wife were members of the Church of Christ, but quickly added that they were not too active, yet were planning on going back to church. I supported this comment with "10-4, that's part of your life that shouldn't be neglected." "Laredo" also responded with a 10-4.

"Why are you going to the 'Superdome City?' " I asked. His response hit me hard. "My son is in the hospital there and has been since November." (This conversation was taking place in August.) I quickly asked what the trouble was, and again the answer was alarming. "They can't seem to find out what the problem is and so can't treat him properly." I detected a note of sadness as his voice came through my unit.

"Please know I'll remember him in prayer," I said, trying to be of some comfort. He said he would appreciate that. Then, with unusual boldness, I felt inclined to make the following transmission: "WOULD YOU LIKE FOR ME TO PRAY FOR YOUR SON RIGHT NOW?"

I heard a woman's voice in the background as Laredo said "10-4, that would be most appreciated." I said, "Let's pray right now for your son."

Heavenly Father, be with Johnny in the hospital. Guide the doctors. Comfort him and be with his family in this time of uncertainty. We know that you are with him and claim for him your blessings. In Jesus' name we pray, Amen.

Laredo's wife was also appreciative of the prayer and concern. Evidently, hers was the woman's voice I had heard earlier in the background. His response was, "Thank you so much–*we* really appreciate it." He quickly said his wife thanked me too.

By this time, I was nearing an intersection, and our paths would part. We said goodbye, and a sense of joy and fulfillment bubbled within me. My thoughts of the recent conversation went through my mind again and again. Because of what had happened, I then realized that God could use a CB radio to proclaim the unsearchable riches of his kingdom. I had bought it as a toy–yet was shown that it could be used to honor and glorify the name of Jesus Christ. I have never forgotten that incident. Nor was I to forget what happened next.

I went to my meeting in Slidell, and the next day was heading to another appointment in Opelousas, actually retracing some of my trip of the day before. As I drove, I continued using my CB, and between Baton Rouge and Livonia I heard a breaker for the Louisiana Canine. I acknowledged and, much to my surprise, it was Laredo going back to his home-20. I immediately inquired about

Johnny, and the report was good. The boy was better and should be coming home in a few weeks. "That's a big 10-4," I said.

Voluntarily Laredo shared again how much he and his wife appreciated the prayer and concern shown the day before. Under my breath I said, "Lord, this was no coincidence. I thank you for helping me to learn this lesson in communication."

The signal became more and more faint as we got further apart, and the conversation ended. But to me, because of what had happened, I knew that my life as a CB'er would never be the same. As of this writing, I've never talked with Laredo again, but there have been others. A word here, a word there, can be used by God to communicate the gospel of Christ.

I hope that you find the incidents and suggestions in this book helpful to you. We are commissioned to share Christ with the whole world—even through the microphones of a Citizens Band radio!

PART ONE

Communicating God's Word Wall to Wall

I have at times tried to imagine what it would be like to live in a world void of all forms of communication. A world without signals, signs, or symbols; a place with no innuendo of gesture or grin, no movement, no proximics. But, like waking up from a dream before a fatal crash occurs, I have never been able to completely imagine such a situation, for the process of imagination itself relies so much upon sensory images which of themselves communicate meaning.

Communication, like the air we breathe, is such a part of our lives, we sometimes fail to appreciate its significance. Yet we know that to remove it *in toto* would be disastrous. Studies have shown that individuals isolated from contact with human beings, sound, and visual stimuli develop bizarre behavior. Hallucinations, disorientation, and even attempts at self-mutilation have been observed. One thing seems clear—people need to communicate.

Some, like Henry David Thoreau, feel they can commune with nature; others describe their experiences with animal life. I, personally, get more satisfaction from talking to a person than to a tree, but the joy of nature cannot be denied. Whatever the medium, the need is still there.

It is this need that has caused man to extend communication beyond the barriers of time and space. In the days before the telephone, the Indian had his smoke signal and his drum. Others used pigeons to carry messages. Today, the telegraph, television, and Telstar communications bring opposite sides of the world together.

Telecommunications predictions for the future seem almost too incredible to believe. Yet, a study prepared by the Arthur D. Little Corporation for the White House Office of Telecommunications Policy envisions that in just the next fifteen years:

- The telephone network will deliver color facsimiles of letters into homes, thus competing with the United States Postal Service for mail service.
- Newspapers and magazines will be electronically printed on television screens.
- AT&T will carry all television service into homes over telephone lines.
- Educational programs, medical information, and defense data will be carried over a government-run communications satellite system.

Today we are seeing the upsurge of perhaps the most popular wireless communications apparatus of all times—the CB radio. Though it was first designed as a network for emergency and necessary business transactions, more and more private individuals are becoming fascinated with its use.

Reports indicate that, at this writing, between 10 million and 15 million sets are in operation in the United States alone. An article in *Science Enterprise News* states that probably more radio operator licenses will be issued in 1976 than the total number of licenses granted since 1912 when the first one was issued.

Some have suggested that CB holds real value for the Christian in his ministry. At first thought, such an idea may seem doubtful. Yet, when we examine the Scriptures, we see that many media have been used to communicate the message of God's love throughout the centuries.

True, in ancient days there were no electronic devices to carry visual and auditory images great distances. There were, though, many signs and symbols used to impart spiritual meaning.

1. Signs and Symbols to Communicate God's Word

The Old Testament contains many references to signs. Some of these are evidence of the miraculous; others are bearers of particular ideas. The rainbow sent by God after the great flood was a sign of his promise to never again destroy the earth by water. The twelve stones described in Joshua 4 represented the twelve tribes of Israel and were given as a reminder of the miraculous stopping of the Jordan River when the Ark of God went across. The blood on the Israelites' doorposts was a sign that their firstborn sons and animals would be spared while the firstborn of the Egyptians would be slain.

Ezekiel's act of openly digging a hole through the wall at Jerusalem and leaving by darkness is an example of an event, rather than an object or substance being used as a sign.

Some occurrences were recognized as signs only by those who were given special understanding. Jesus regarded himself as the sign of God's redemptive activity and resented the attitude of those who wanted him to give a sign to prove he was the Son of God (He even rebuked Satan's challenge, "If thou be the Son of God, command these stones be made bread.") However, he did describe certain signs which would alert his followers to the nearing of the second coming. Mark 13:4-37 records these as wars, earthquakes, famines, persecutions of the faithful, and preaching of the gospel to all nations.

Whereas signs were essentially conveyors of revelation or warning, symbols were actually a part of the language of faith–words, objects, names and places embodied meaning which helped describe and mold conduct and belief.

To the ancient Hebrew, words were more than auditory or visual symbols. They were reality. They defined and determined life. They related more than a message; they carved the course of lives.

Words were spoken seriously and taken seriously. One did not bless or curse frivolously, for the consequences were great. Esau's life took a sore turn because his father mistakenly blessed Jacob, his brother. Though Isaac trembled to learn his error, the blessing could not be taken back.

God's word was final. (Isa. 45:23, TLB) ". . . so also is my word. I send it out and it always produces fruit. It shall accomplish all I want it to, and prosper everywhere I send it." Judah felt the consequences of God's curse for their sins. God said they would be left desolate, destroyed by drought and . . . they were.

In those days words were so tied to experience that even names were chosen to carry meaning. In fact, God himself ordered that some persons should be given specific names. Abraham (father of the nations), Maher-shalal-hash-bag (your enemies will soon be destroyed), Shear-jashub (a remnant shall return), Lo-ammi (not my people), Peter (the rock), and Immanuel (God with us) were names assigned to convey specific messages.

The ultimate word came in the person of Jesus Christ. He was the Word made flesh; the culmination of God in human form.

The Deity of God was expressed symbolically.

> "God could be portrayed visually; hence he is described symbolically, so as to make comprehensible his nature and activity. He is represented with features of a man, a beast, or nature itself. His eye devotes his omniscience . . . and favor . . . ; his arm, his power . . . ; and his right hand, victory, vindication . . . He comes in judgment like a roaring lion . . . , an East wind, a storm . . . By all such descriptions God was made intelligible; his will and purposes made plain; and his activity, as a living force of the vicissitudes of life made known."[9]

Objects communicated meaning. The symbolism of fire can be seen in these verses: "But who may abide the day of his coming? and who shall stand when he appeareth? for he is like a refiner's fire . . ." (Mal. 3:2). In this case, fire is symbolic of purification. In the verse which follows, it is representative of God's glory. "Those at the bottom of the mountain saw the awesome sight: the glory of the Lord on the mountain top looked like a raging fire" (Ex. 24:17, TLB). Fire too denoted the wrath of God: "His anger flared out against them because of their complaints, so the fire of the Lord began destroying those at the far end of the camp" (Num. 11:1, TLB).

Parts of the body were also symbolic. For example, the heart signified intelligence and the eye, mental perception. Deuteronomy 12:23 (TLB) shows that the blood represented life: ". . . for the blood is the life, and you shall not eat the life with the meat."

Perhaps the most beautiful and meaningful symbolism in the Scriptures centers around the person of Christ. Those who shared the message and mission of Jesus used figurative terminology so that all could understand.

Paul referred to him as the peacemaker, the Lord of glory, the only foundation, the image of God, the high priest, and the author of our faith. Isaiah portrayed him as Immanuel, the mighty God, the everlasting Father, the prince of peace. In contrast to these more glorious sounding titles, he called him servant. Such contrasts seem contradictory. Yet we know that Christ lived the message of service he preached.

One of the most poignant portrayals of Christ is that of the lamb of God who takes away the sins of the world. This symbolism carries with it the significance of Christ's sacrificial death. How paradoxical that this lamb which was slain was also the Messiah (John 1:41). How humbling that this Son of man is also the Son of God, the King of kings and Lord of lords.

2. The Greatest Communicator of All Times

Sometimes messages are conveyed by the most unlikely messengers. When Beau and I first married, we were privileged to know a delightful, beautiful teenager named Cindy. Cindy was the sort of person everyone enjoyed being around. She had one of those effervescent personalities that would have made a fortune if it could have been bottled and sold.

However, what she had in personality, she lacked in brains. No one would ever have accused Cindy of being smart. In fact, part of her charm lay in her naïvete and the childlike way she expressed herself. Yet one Sunday morning, this teenager, whom most considered void of all depth, surprised a roomful of listeners by relating the following story. She began by saying, "I don't know how to put things very well, but maybe this story will help you understand what God did for me."

"A number of years ago a man went to work as a bridge operator. It was his job to open and close a section of railroad tracks to allow trains to cross over a big bay. The man had a four-year-old son named Will who was his pride and joy. Because the father's work involved a lot of inactivity and tended to be a lonely occupation, he let

Will join him at the station house from time to time.

"The child was fascinating to watch. No roley-poley bug or velvety butterfly escaped his eye. His tiny hands built everything from dinosaurs to battle stations with the sand that collected outside the station house. The trains were his friends, and he shouted happily each time one passed by. Whatever he did delighted his father and helped him pass the long hours.

"Will's father coached him very carefully to avoid playing near the tracks and to avoid, particularly, a section near the bridge which was obscured by trees. Yet, one day, Will became involved in chasing a mosquito hawk and wandered along the tracks out of the earshot of his father.

"Near the time a passenger train was due, the bridge operator noticed his son was missing. He began searching frantically for the boy, shouting Will's name until his throat failed to choke out another word. He knew he had to get back to the station for the bridge was open, and the train was scheduled any moment. As he was approaching the control booth, he spotted Will. Cold sweat broke out all over his body. There was Will directly in the path of the train and hidden from its view!

"Only one escape was possible, but it involved the father leaving the bridge open and allowing all on the train to plunge to their deaths. There were only seconds to make a decision. In those seconds he agonized fitfully, 'My son or a trainload of strangers; the joy of my life or . . .'

"As the train passed by, he sank to the floor emptied of all strength. He wondered if any of the people aboard, laughing and talking, were even aware of the sacrifice that had been made for them."

As Cindy spoke, I thought, "What if it had been my child? Could I have done this?" I concluded I could not have. Yet, that is exactly what God did for us. He allowed his only son to be sacrificed so that we might live.

I am thankful God used Cindy to vivify the meaning of God's gift to me. Before the narrative, I had never stopped to think that God could have chosen not to throw the switch. The decision was entirely up to him. He chose to let me live.

I have included this story as a means of telling you why I consider Christ the greatest communicator of all times. I make no claim to be a Bible scholar or a communications expert; I can only tell you why a man who lived over 2,000 years ago still communicates with me today.

In my way of thinking, other "greats" of history aren't even in the same league with Christ. I do not argue the hypnotic power of Hitler as he shouted his message of madness to the masses. Nor do I belittle the influence of scholars like Plato and Aristotle. Certainly the beauty communicated by Michaelangelo's brush and chisel cannot be denied. The achievements of Shakespeare are still unsurpassed by those who compete in his field.

But none of these has ever done what Christ has. He himself *is* the message and the mission he preached. Perhaps Dwight L. Moody states it best with this illustration from his sermon "What Christ Is to Us."

> "I asked a man sometime ago why he thought so much about a certain man. I noticed that he could not speak of him but tears came into his eyes, and so I asked him, 'Why is it that you love this man as you do?' 'Why, Mr. Moody,' he said, 'that man saved me.' "

Therein lies the difference. The messages of all others stop short. They only *tell* about life. Christ *gives* life. This fact, of course, speaks of the divine nature of Christ, and no man should even hope to compete with such supremacy. But Christ was more than divine. He was a man, and he communicated as men do. As followers who would proclaim the gospel, there is much we can learn from Jesus' brand of communication.

As I consider the way Jesus spoke, there are certain things that impress me. First of all *Jesus spoke simply.* He used pictorial language which could be understood by people from all walks of life. He talked about things that were familiar—bread, wine, water, salt, fields, and light were all images he used to express religious truth. In proclaiming the message of salvation he drew up scenes from the marketplace, the law courts, the Temple services, the cultic groups, and the family circle. In essence, he related spiritual truth to the daily experiences of those to whom he spoke.

Parables were an integral part of his messages. Matthew pointed out that Jesus certainly used these illustrations when speaking to the crowds. In fact, because the prophets said that he would use so many, he never spoke to them without at least one illustration. For it had been prophesied, "I will talk in parables; I will explain mysteries hidden since the beginning of time (Matt. 13:35, TLB).

Scharlemann says, "The word parable belongs to the language of revelation, like 'grace' and 'faith' it belongs to the lexicon of 'good news.' It is descriptive of God's activity rather than being either prescriptive or predictive." [12]

Four different types of parables have been distinguished by various writers: those which paint a picture of some aspect of God's kingdom, those which are allegorical, those which make comparisons, and those which are simple sayings. Because of the overlapping of some of these categories, estimates of the number of parables in the Gospels vary from thirty (Trench) to seventy-nine (Von Keetsveld). Whatever the number and type, they accomplished what they were designed to:

> Then he explained to them that only they were permitted to understand about the Kingdom of Heaven, and others were not. "For to him who has will more be given," he told them, "and he will have great plenty; but from him who has not, even the little he has will be taken

away. That is why I use these illustrations, so people will hear and see, but not understand" (Matt. 13:11-13, TLB).

In all his ministry, *he identified* with individuals. "The crowds heard him and He wanted them to have what he came to give, but he knew full well that the really important things in the world happen within the souls of individual men." [13]

He was not the high-faluting, better-than-thou sort of teacher who felt himself too good to be seen with those whose halos were a little tarnished—quite the contrary. One reason he aroused so much controversy among the Jewish uppercrust was his insistence on reaching out to swindlers, robbers, and others of questionable character. He paid no attention to those who criticized him. "The Word was made flesh, and dwelt among us" (John 1:14).

When he and his disciples dined at Matthew's house with a group of swindlers, the Pharisees were indignant, but Jesus shamed them by sending them away with instructions to learn the meaning of the Scripture. "It isn't your sacrifices and gifts I want—I want you to be merciful" (Matt. 9:13, TLB) (Wouldn't that have been fun to see—Jesus' saying, "Run along now, boys, you've still got some more homework to do"?) He made it perfectly clear that he had come "not come to call the righteous, but sinners to repentance" (Matt. 9:13).

Even his own disciples were shocked to see him talking to the Samaritan woman who had gone to draw water. Jews just *didn't* speak to Samaritans.

But there was Jesus talking, not to just a Samaritan, but to *a Samaritan woman* who had five husbands and wasn't even married to the man she was living with! What *would* the Joneses think?

Well, some pious leaders caught a woman in the act of adultery, placed her out in front of a staring crowd to be ridiculed. They had heard how Jesus treated sinners and were anxious to see if they could get him to say something they could hold against him. He spoiled all their fun by telling them to go ahead and stone her to death, as long as the first stone was thrown by someone who had never sinned!

In her book, *The Hiding Place,* Corrie ten Boom describes her and her sister, Betsie's, reactions to life in a German prison camp during World War II. She testifies in regard to one incident, "I glanced at the matron . . . I saw a gray uniform and a visored hat; Betsie saw a wounded human being." [15] The Pharisees looked at the adulteress and saw a Mosaic law that had been broken; Jesus saw a soul that needed mending.

Isn't it marvelous that this Lord of compassion and forgiveness is also a Lord of authority? He recognized his authority, and he spoke with it. "I, the Messiah have the authority on earth to forgive sins. But talk is cheap–anybody can say that. So I'll prove it to you by healing this man" (Matt. 9:5-6, TLB).

Webster's Dictionary defines authority as "the power or right to give commands, enforce obedience, take action, or give final jurisdiction." Jesus possessed all these. His power source was God himself. No other authority was needed. His disciples knew his supremacy. "You call me 'Master' and 'Lord' and you do well to say it, for it is true" (John 13:13, TLB). Even the winds and the seas obeyed his voice. Though he was, of all men, most humble, where God's will was concerned he was unwavering.

In the Temple, he gave no thought to popular opinion. His fury turned against those who defiled God's house by turning it into a marketplace. Without hesitation, he knocked down their trading tables and disrupted their negotiations. He saw to it that only God's business went on in God's house.

When confronted with opposition he dealt with it quickly and effectively. Certain Jewish teachers accused him of being possessed by Satan because he cast out demons. He confronted them and asked, "How can Satan cast out Satan? A kingdom divided against itself will fall . . . and if Satan is fighting against himself, how can he accomplish anything?" He was *committed* to *communicate* God's word at all costs.

St. Augustine declared, "but whatsoever may be the majesty of style, the life of the speaker will count for more in securing the hearer's compliance."[5] Jesus' style was superb. He spoke simply and with authority; he related to individuals; he lived the message he shared.

3. Guidelines for Christian Communication

First Timothy 4:12 states clearly, ". . . be thou an example . . . in word, in conversation, in charity, in spirit, in faith, in purity."

George Miller, author of *Language and Communication,* writes that human vocalization is the most important encoding procedure of all communication. He claims that millions of words are in existence and that we could spend weeks of continuous filibustering without using the same word twice. That words play an integral role in human communication is dramatized by Alvin Toffler in his *Future Shock.* He points out that between 10,000 and 20,000 edited words per day are consumed by the American adult and that "the same person also probably spends an hour and a quarter per day listening to the radio–more if he owns an FM receiver. . . . If he listens to news, commercials, commentary, or other such programs, he will during this period hear about 11,000 preprocessed words. He also spends several hours watching television. Add another 10,000 words or so, plus a sequence of carefully arranged, highly purposive visuals." [16]

Toffler adds to these a minimum of 560 advertising messages giving an outstanding impression of how thoroughly bombarded the average American adult is with words. So vital are words to our way of life that "Super-industrial production requires workers skilled in symbol manipulation . . ."[16]

Certainly words are part of our way of life. For society as a whole they reflect trends in economics, politics, technology, family life. For the individual citizen they are a mirror of what we are within.

"For a man's heart determines his speech. A good man's speech reveals the rich treasures within him. An evil-hearted man is filled with venom, and his speech reveals it" (Matt. 12:34-37, TLB).

I can think of no better way to describe the kinds of words we should speak than to cite the Scriptures themselves: "Let no corrupt communication proceed out of your mouth, but that which is good to the use of edifying, that it may minister grace unto the hearers" (Eph. 4:29). "Don't ever forget that it is best to listen much, speak little, and not become angry, for anger doesn't make us good, as God demanded we must be" (James 1:19, TLB).

"Thou shalt not take the name of the Lord thy God in vain" (Ex. 20:7). "Swear not at all; neither by heaven; for it is God's throne: nor by earth; for it is his footstool: neither by Jerusalem; for it is the city of the great King. Neither shalt thou swear by thy head, because thou canst not make one hair white or black. But let your communication be, Yea, yea; Nay, nay: for whatsoever is more than these cometh of evil" (Matt. 5:34).

"Keep thy tongue from speaking evil, and thy

lips from speaking guile" (Ps. 34:13). "He that speaketh truth sheweth forth righteousness" (Prov. 12:17). "Lying lips are abomination to the Lord . . ." (Prov. 12:22). "Make them pure and holy through teaching them your words of truth" (John 17:17, TLB). "Likewise must the deacons be grave, not double-tongued . . ." (1 Tim. 3:8).

"A soft answer turneth away wrath" (Prov. 15:1). "A word fitly spoken is like apples of gold in pictures of silver" (Prov. 25:11). "If any man offend not in word, the same is a perfect man, and able also to bridle the whole body" (Jas. 3:2). "But as for you, speak up for the right living that goes along with true Christianity" (Titus 2:1). "Whoso keepeth his mouth and his tongue keepeth his soul from troubles" (Prov. 21:23).

"Be patient and you will finally win, for a soft tongue can break hard bones" (Prov. 25:15, TLB). "Anyone who says he is a Christian but doesn't control his sharp tongue is just fooling himself, and his religion isn't worth much" (Jas. 1:26, TLB).

From these Scriptures, and others not mentioned, it can be seen that truth and appropriateness of our words are necessary; our attitudes and emotions must be kept in check; swearing and cursing must be avoided. These guidelines are important to us as we attempt to be *good examples in our conversation.* That is, conversation in the Scriptural sense of the word.

While the modern interpretation of conversation refers to verbal communication or colloquy, the ancient sense of the word is much broader. It includes one's behavior and conduct. In that regard we are told to "Let your conversation be gracious as well as sensible, for then you will have the right answer for everyone" (Col. 4:6, TLB).

Some time ago I had occasion to attend a professional conference where a very brilliant man was demonstrating a therapeutic technique. As I observed this man, who is recognized internationally as one of the leaders in his field, I was astounded by the way he controlled his patients' every mood. Even the most hostile patients ended up searching for ways to please him. His office was filled with mementos and gifts that people had given him. His patients absolutely adored him. One woman I happened to catch alone for a few minutes said, "I'd do anything for that man. He's the dearest person I know."

What made him so special? My curiosity was killing me, but I concluded I would never know. I assumed I would never see him again after the conference terminated. Surprisingly, a series of unexpected events led to my knowing him personally. After two years of professional association with him, I understand why his patients are so devoted to him. He knocks himself out to serve them.

During the times that I have been with Dan, I have seen him at 7:15 A.M., sharing donuts with the janitor at the exclusive office complex where he works; at 8:30 A.M. checking on a conference participant who was reported to be ill; and at 9:30 P.M., putting together materials to send a frustrated beginning therapist. I have seen him take time out from his lectures to get an assembly of doctors from all over the world to sign a sympathy card for a participant who was called away because of his father's death. I have discovered him giving up his lunch hour to help a student stranded without funds. The list of "little things that meant a lot" is unending.

Dan's concern for each individual is genuine. He makes each person feel he is of worth and importance. He speaks quietly and with wisdom—never criticizing, never forcing. In short, he lets his actions and words communicate a love for others. I can't imagine him ever intentionally doing anything to damage another person's self-concept. We should hope that we as Christians "never do anything that will cause [us] to be ashamed of [ourselves] but . . . to always be ready to speak out boldly for Christ" (Phil. 1:20, TLB).

The Pharisees of old had scores of laws designed to govern their behavior, and they painfully heeded them. Jesus told them their laws did them no good as long as their hearts were evil and cold.

He showed them that their *conversation* (conduct, deportment) *should be based on a charitable heart* which is pure and right in spirit. What then does it mean to have a charitable heart?

The word charity was created out of the life of Christ and appeared first in the apostolic writings.

Unlike the current concept of charity as almsgiving, the original term *charis* was closely akin to agape (love). As described in the Pauline writings it is "the motive that determines the quality of the whole inner life and gives values to activities . . . It determines and defines a man's own character and personality. It is not boastful or arrogant, but dignified, pure, holy, serene. Evil cannot provoke it nor wrong delight it."[8]

In my opinion, the Scriptures' most beautiful treatment of love is found in 1 Corinthians 13. To try to improve on it would be an exercise in futility; to attempt to explain it would be demeaning. It is self-explanatory. Let me mention a few whose lives have exemplified the characteristics described in those verses:

> From the security and warmth of a close-knit family in Holland to the horror and stench of a Nazi prison, Corrie ten Boom and her sister, Betsie, carried a love that transcended human love. It was a love (charity) that reached out to persecutors, as well as friends, and offered forgiveness for abuses too terrible to describe.
>
> From farmboy to fame in the major leagues, Billy Sunday could not escape the call of Christ. Once surrendered, he seemed driven to spend every moment proclaiming God's word. His methods were criticized by some. He was described as undignified by others. Whatever his style, his love was genuine. It led him to give

up a fortune in baseball to carry the message of salvation from one end of the country to the other. He pushed himself to the very point of collapse. Some said that he spoke like a "dying man to dying men."

From a child angered and confused by the barriers of blindness, deafness, and muteness, Helen Keller became a miracle in her own time. The love and patience of her teacher, Miss Sullivan, overcame the impossibility of breaking those barriers.

Others who have exemplified this ideal and principle of life might be mentioned. Martin Luther, Albert Schweitzer, Billy Graham, William Wallace, and others. When I study the lives of such people, I am made aware that more than human discipline and strength was required to help them accomplish their missions of love. *Faith* was necessary. Neither the faith that bespeaks simple optimism, nor that which accepts something without reason, but faith which is reliance upon God.

In the Scriptures, faith has other meanings, but it is most often used in the context of reliance or trust. Hebrews 11 says it is the "substance of things hoped for, the evidence of things not seen." Moses, Abraham, Elijah, Paul and others relied on God for strength and direction. Without such faith they could never have achieved the incredible things they did for "without faith [reliance] it is impossible to please him: for he that cometh to God must believe that he is, and that he is the rewarder of them that diligently seek him" (Heb. 11:6).

In the sense of "belief" or "creed," James 2 stresses the futility of faith without works. Verse 22 uses the word faith to mean both reliance and creed. "You see, he [Abraham] was trusting God so much that he was willing to do whatever God told him to; his faith was made complete by what he did, by his actions, his good deeds" (TLB).

Trust in God is necessary if we are to *communicate in the right spirit.*

God's Word shows us there are some spiritual characteristics to be encouraged and others to be avoided. Humility (Matt. 5:3), gentleness (Gal. 6:1), and patience (Gal. 5:22) are qualities to be desired. A spirit of wisdom, understanding, counsel, knowledge, and fear of the Lord (Isa. 11:2) are to be treasured. If we fail in achieving a right relationship with God, a spirit of "whoredom" (Hos. 4:12) prevails. A right relationship may be broken by a spirit of jealousy (Num. 5:14), or pride (Eccl. 7:8), of haste (Eccl. 7:9), and of rebellion (Ps. 78:8).

One knows instinctively whether his spirit is right for "a man's conscience (spirit) is the Lord's search-light exposing his hidden motives" (Prov. 20:27, TLB). In this regard we are advised to exercise control over our own spirits (Prov. 16:32).

All our *communication should proceed from a heart that is pure.* Christ said that impurities from within us are the things that corrupt us. "There is nothing from without, that entering into him can defile him: but the things which come out of him, those are they that defile the man" (Mark 7:15). Our attitudes and thoughts significantly influence our behavior. If we aim to impress others favorably as we communicate with them, it would be wise to follow the instructions in Philippians 4:5-9:

> "Let your moderation be known unto all men. Finally, brethren, whatsoever things are honest, whatsoever things are just, whatsoever things are pure, whatsoever things are lovely, whatsoever things are of good report; if there be any virtue, and if there be any praise, think on these things. Those things which ye have both learned, and received, and heard, and seen in me, do."

We can never attain the level of perfection that Christ did in transmitting the Gospel, but if we strive daily to exemplify him in word, conversation, charity, faith, spirit, and purity, we open up the way for God to use us.

PART TWO

Ways the Christian Can Use CB

In preparing this book I took a survey from CB'ers all across the United States. I was amazed to discover how many persons have found CB modulating to be a tool for witnessing. I was not as impressed with this fact as I was with the number of people who fall into this category and with the different ways in which CB's are used in Christian-related tasks.

As you read of these examples and testimonies from various individuals, no doubt you will recognize similar circumstances in which you have found yourself. Also, I'm sure the wheels will begin to turn in your own mind as you think how you might adapt and put into action some tasks of like design.

Christians can use a CB radio in many ways. In fact, the potential is almost limitless. One of the most popular ways has been picked up by churches and, more specifically, church staffs.

4. Visitation Ministry

More and more churches are putting CB units in staff members' cars so the staff can be in touch with one another. Some churches are also putting base stations in the church office. If an address is needed or an urgent message needs to be forwarded to the pastor or staff member, the secretary can make contact and provide the information.

First Baptist Church of Wichita Falls, Texas, uses a two-way radio for its staff. This is more sophisticated (and more expensive) than CB, but offers a private channel and longer range. However, the CB setup can be just as effective and, of course, is not as costly.

A pastor of an East Texas church says that all of their staff members have CB's and are in touch with each other at a moment's notice. He related that a teenage boy was drowned near his community, and the word came over the CB to try and locate the father. The pastor put out a 10-5 and was able to locate him.

The pastors in large metropolitan areas (and even small towns) can use CB effectively in the following manner. Usually, there are a number of base stations that are monitored almost 24 hours a day. I'm not speaking of the emergency Channel (9), but of individuals who are caught up in the hobby and are 10-10 (listening) most of the time. It will benefit a pastor to know some of the base station operators, and frequently modulate with them. Then, when an address is needed while out visiting and he doesn't have time to go back to the church, or it is too far, then he can rely on such operators to make a "land line" call for him and secure the information.

You'll probably be surprised at the number of persons in your own church who have CB units and even base stations. Most are more than willing to help out the pastor on occasion. Cultivating relationships like this can also lead to witnessing and counseling opportunities, like the one experienced by Calvin Cantrell (The Padre), associate in the evangelism office of the Louisiana Baptist Convention. This happened to Calvin several years ago while he was pastoring a church in North Louisiana.

Calvin testifies: "The guy is dead now, but this even adds to the story. He was director of recreation for the City of Monroe and in charge of a big city gymnasium, fall leagues, etc. He had sugar diabetes which eventually got to his eyes. The city let him keep his job for months and months–even after he lost his sight completely.

"We both had good base stations, and I can remember many nights talking into the late hours with him. My unit was right by my bed and very convenient. One night I was listening and heard my friend modulating with several 'good buddies.' I could tell his stinger was out and he was mad at the world about something. I broke in during a pause and scolded him. I told him he sounded like he was blaming God and everybody else for his troubles. And he said, 'Well, I am.' This statement upset me, for I knew then that he must have some real problems. *I didn't know* at the time he was blind, but I began to tune in a little more sensitively because of what he had said.

"I first suspected something physically wrong when I heard him answer folks about his S/RF meter (a meter used to measure strength of incoming signals and transmitter power). He would guess at the numbers. 'What am I putting on you?' they would ask. He would reply, 'seven, six, or maybe even five.' He was faking and wasn't letting anybody know he was blind, for it embarrassed him.

"By and by as time went on, he and I made friends on the radio, and it turned out that my scolding him was just what he needed. Other people were sympathizing with him or pitying him, but he didn't want pity. He wanted somebody to be firm with him and help him find his way. I was the only one on the channel that would be firm with him, having verbally 'knocked the fire out of him' about his terrible attitude. I would have been just like the rest of them if I had known he was blind, but I didn't. I was just being me and letting him be himself.

"He would ask me questions about the Bible, and I would answer as best I could, using terms that were not 'preacher words.' It was an interesting conversation, and several people would tell me later that they heard us talking. Later on, I made an appointment to go and visit with him. When I walked in the door I was almost speechless. The fact that he was blind caught me off guard. Yet, that very day I led him to the Lord in his living room. He lived way across town from the church where I was pastor–a long way, but his wife would drive him out to the church every Sunday. He united with our church, I baptized him, and he was very faithful for a period of about four months. A couple of weeks after that he entered the hospital and died."

"The Padre" continued with this testimony, which directly relates to the use of CB in visitation ministry.

"A few years ago when there weren't so many people on the channel, I had a CB radio base at my house and one in my car. My wife was the church secretary and when she needed me for some reason or someone in the church needed the pastor, she would go next door to the parsonage and contact me over the air. I had the CB on PA (external speaker) and if I was in a house visiting or at the hospital, I could hear a call through this device.

"Sometimes I would get a 10-5 (relay) from a friend and he would give me the message to call the office or follow up on whatever the message was. People all over town were tremendously help-

friend and he would give me the message to call the office or follow up on whate-er the message was. People all over town were tremendously helpful. The unique thing about this was that the entire town shared in my ministry. It helped the people to know that there was somebody on call who really cared–someone who would answer a call day or night."

Richard Sims, consultant for the Texas Baptist Sunday School Division, uses his CB while visiting in and working with churches around his state. His handle is "The Counselor," and oftentimes he is able to talk with persons on the road. He talked with one who wanted to know what Christians believe and what church he might attend in that particular area. Sims mainly uses his for locating churches and pastors in various towns. He, too, uses base stations advantageously.

Another pastor relates how he became sold on CB. He was holding a revival in a rural area and as he visited with the resident pastor, he was exposed to CB use. Practically all of the farmers in that area had units. He said, "That's the best revival promotion I've seen. Messages (reminders) about revival services were relayed all over the countryside. We had a great meeting!"

W. D. Carson, pastor of the Pinehurst Baptist Church, Ball, Louisiana, uses his CB frequently in his pastoral ministry. Several of the members in his church have base stations and even more have mobile units. Brother Carson has had some unique experiences in CB which have enabled him to minister to many different types of people. Here is one such experience:

"I modulated with a young man quite a lot over

a number of months, and our friendship began to grow. He asked if I would come by and talk with him, allowing an opportunity to ask some questions that had been bugging him. As we talked, I found that he was hung up on getting baptized. He mentioned that every other preacher he talked with made an issue of it. We spent over three hours going over the meaning of baptism and how the Scriptures pointed to all the answers.

"The following Sunday was Easter, and he united with our church on profession of faith in Christ, influencing the entire family in the Christian faith. That night, he followed Christ in baptism. Had it not been for my CB radio, I am almost positive that this incident would never have occurred."

Brother Carson is also active in REACT (Radio Emergency Associated Citizens Teams), serving as chaplain of the Central Louisiana unit. I asked him why he wanted to be a part of this group, and his reply was: "There's a certain romance about keying a mike that brings people together. REACT and other similar groups want to minister to people, and that's what Christianity is all about. I feel I can help them to minister in Christ's name by working right beside them."

Carson went on to say, "In the last year, eight persons have united with our church on profession of faith as a direct result of a CB radio. In personal visitation, contacting members, and community service, I find a CB most valuable."

5. Bus Ministry

A number of churches across the country are utilizing CB in their bus ministry. One church has been using CB for almost five years. It should be noted here that units are not only convenient and helpful, but are good for safety reasons.

The Chowchilla, California, incident, in which a bus load of children were kidnapped, promoted officials in that county to install CB units in all buses. Not only can the driver use the unit to find directions and offer assistance, but he can also summon help should it be needed.

When I was on the staff of a church in North Central Texas, we ran ten buses on Sunday. I can still remember going after children whose parents called in saying their kids missed the bus. Often, the bus would only be a few blocks from that address, but there was no way to contact the bus driver. Churches with small and large bus ministries are finding CB to be most helpful in rounding up bus riders.

With the advent of the "trip bus," churches are installing units for convenience and safety. Breakdowns are usually more frequent on the older buses, and many churches are still using older models (out of necessity) to make short and long

CB can be useful to those churches involved in bus ministry.

trips. Being able to find out road conditions and hazards ahead gives peace of mind to driver, passengers and parents back home.

I have driven an "over-the-hill" bus many a mile and have had several breakdowns. But at that time, I didn't even know that CB radios existed. So I feel that any bus ministry can be strengthened significantly by using the "squawk box."

6. Mission Trips

This particular ministry is strongly associated with the bus ministry, because of the vehicles involved. Most church groups (youth groups, choirs, disaster units, dental mobile units, medical units, etc.) will bus persons into an area. Also, most mission groups will travel together in cars or a combination "convoy" of bus(es) and cars. CB units can be most helpful to help you stay in touch on the trip in the event of breakdowns and other contingencies. But more importantly, the units are useful in the mission field.

Many youth groups will "blitz" an area for perhaps a week or so, having a "people search" in the mornings, Vacation Bible School in the afternoons, and even revival services at night. Walkie-talkies, which are really one-channel CB units (for all practical purposes), can be functional during these activities. Coordination is most important, especially for the "people search." Mission areas are often low-income housing developments and many times streets are not named or numbered, making it hard to keep from duplicating a section or missing one entirely.

Gene Pepiton, minister of education at the Pineville Park Baptist Church in Pineville, Louisiana, mentioned that members of a mission team from his church stayed in touch with one another as they traveled to and from mission points in a Western state where they were holding Vacation Bible Schools. And, of course, they reported on the events of the day and shared testimonies about what happened in their schools.

Walkie-talkies, mobile CB units, and units with P.A.'s also provide a drawing card to youngsters in mission areas. They are fascinated by the "talking boxes" and the amplification. Drawing a crowd like this will help to spread the word that something different is happening in the neighborhood, and "I don't want to miss it."

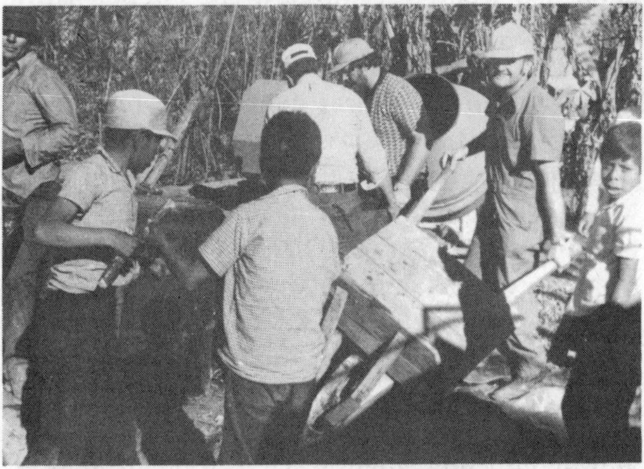

John Winters and other volunteers assist in relief operations in Honduras following the 1974 earthquake (Foreign Mission Board Photo).

7. Disaster Relief

The Texas Baptist Men's Department in Dallas has utilized CB units very effectively, especially in disaster relief. Usually, power is out in disaster areas, and workers can stay in constant touch with each other and other disaster workers. Supplies can be requested, assistance given, and information shared.

John Winters, disaster relief director for Louisiana Baptists, related the following incident that happened while directing a disaster relief trip in Honduras.

"I am sold on CB's and Hams. A group of us were down in Honduras doing some relief work after the terrible hurricane a couple of years ago. The eighty-year-old mother of one of our workers was involved in an accident in which her leg was broken in four places. There were no telephone communications into Honduras."

He continued, "I had left information with my wife to contact a Ham operator in Longview, Texas, in the event of an emergency. He had been communicating messages to relief workers in Honduras, and we hoped he might reach us if the need arose. My wife, learning of the accident, called the Ham operator to see if he could make contact. The Ham did make contact with another Ham, and he in turn sent out the urgent message to CB units. Because the worker was out on a job several miles from the nearest large town, a CB operator had to go to him and personally deliver the message.

"The worker was able to catch the only flight out of Honduras that day (Tuesday). There were no flights on Wednesday. He flew into New Orleans and made a 35-minute connection which was a miracle in itself. Ordinarily it takes over an hour just to go through customs. After flying into Alexandria, he was at his mother's bedside at 8:30 P.M.—just eight hours after the accident had occurred. Without the assistance of the two-way radios and CB's, it would have been days before the man could even have been located."

I believe what we're talking about here is *ministry* aided by *communication.* Outreach visitation, bus ministry, missions trips, disaster relief, and other projects are ministry actions supported and aided by the use of the Citizens Band radio. Sure, in most of these cases, CB is not an absolute necessity, but God is blessing abundantly, so why not do everything in our power to help the Holy Spirit work in various situations?

A spirit of cooperation also surfaces when you speak of ministry and mission action. The following testimony from J. D. Scott, retired pastor in Central Louisiana, helps to illustrate cooperation. I'd like to call it *Communicating a Cooperation.*

"I held a revival in the Big Cane community

(Louisiana) several years ago. As I was recently driving through that area, modulating with some folks, a break came in–'break, break, KR6977.' I came back and said, '10-4, go ahead.' He said, 'You wouldn't be a Baptist preacher, would you?' I said, 'Yes, I am.' He continued and asked if I were Brother J. D. Scott. I confirmed that with another 10-4. He went on to say that he had recognized my voice and that I had baptized him some twelve years ago in that revival meeting. Imagine that–his recognizing my voice after that long a time. We talked for a good long while. When we bid farewells to each other just outside of Bunkie, the local Catholic priest came in and said, 'Break for Reverend Scott.'

"I told him to go ahead, and he said, 'I enjoyed the conversation between you and my friend 'Buzz' [The person Scott had baptized] What's your 10-20?' At that time I was about three miles north of Bunkie. The priest went on to say he was trying to catch me before I got out of town because he wanted to talk. 'Well, I'm already out of town and running late,' I conveyed. Then the next statements really surprised me: 'I have been admiring what Baptists are doing–so many things. I hope to cultivate someone who would feel free to talk to me and give me the answers to several things that I'd like to know about–that I care about. I'd like to hear about it from the horse's mouth.'

"I told the priest, 'Well, I've been associated with Louisiana Baptists for about twenty years, and we have no secrets. Anything that we have that will bless the world, I'll be glad to share with you.'

"The priest said, 'I'll tell you what. I get up to Alexandria quite often. Could I meet you sometime, have coffee, and talk with you?' I said, 'Yes sir, anytime. You just call me.'

"A week later he did call, and we got together at a restaurant. The priest wasted no time in getting into the conversation. 'Now, I've been puzzled about a lot of things. I understand that you Baptists have a very extensive foreign mission program. Our missionaries let us know about that. And that your educational program is also very strong.' I said, 'Yes sir, we're rather proud of what we've accomplished through God's leadership.' Then, he asked me to tell him all about it.

"I told him that our tithes and offerings of church members help to make up our missions giving. 'How much do you assess your members?' he asked. Quickly I said, 'It's all volunteer–churches decide what they want to contribute–all voluntary cooperation.'

" 'Well, about how much a year would that run?' he asked, 'a million, two million?' I said, 'roughly a million dollars a week.' The priest couldn't believe what he was hearing. 'You mean you do that without assessing members? Fantastic! Fantastic!'

"I gave him some of the latest printed information on our Cooperative Program. I'm so thankful that the contact took place on the CB. Otherwise, it might never have happened."

Other ways that CB can be used by the Christians and Christian groups are at . . .

8. Retreats and Camps

Jerry Bob Taylor, state Royal Ambassador director for Texas, tells of a camping incident in which a young boy was injured. The boy was taken to the hospital and treated, but on the way back to camp, the doctor called on the CB to tell that he had failed to administer a tetanus shot. They turned around and went back.

In addition to the mobile CB's, the Texas RA Department also uses portable battery-operated units to keep in touch with the leaders of its canoeing and wilderness camp programs.

When I was doing youth work in a local church in Texas, I used intercoms on a limited basis to stay in touch with counselors. CB walkie-talkies would have been more useful.

Family camping outings are becoming increasingly popular. Hundreds of churches now sponsor "Family Camping Retreats" with families pulling campers, driving mobile homes, or packing tents. The CB's provide a convenient way for the convoys which are formed to communicate. Additionally, the units give an extra feeling of security and togetherness to the family groups as they travel toward their destination.

9. Resort Ministry

"Jimmy Brossette, tell me how your CB is helpful in your work as a missionary in a resort area," I asked. He then conveyed these thoughts, "My handle is 'Lake Missionary,' and immediately people recognize from this that I'm there to help. I couldn't do without my CB. I have used it in the following ways:

"(1) Search and rescue
(2) 10-33 (Wreck–Accident)
(3) Fires (forest and auto)
(4) Weather report to fishermen
(5) Telling folks the time of services or the directions to a church
(6) And I always sign off with a 'God Bless You.'

I try to use my CB in all the above ways, and it's all for the glory of God."

Jimmy Brossette has a monumental task in his resort area. He is the missionary for the Toledo Bend Baptist Center. Toledo Bend Reservoir is one of the most famous bass fishing areas in the world. People come from all over the United States to try their luck at pulling the "lunkers" out. The "Lake Missionary" is there to assist. (But not with the fishing.) Give him a "shout, shout" when you're over that way–he'll appreciate it.

Coincidentally, the first director of the Toledo Bend Center also had a CB and was involved in search and rescue. I asked him to relate this story.

"I had a portable rig (CB) that I carried in my boat for emergency use and for monitoring the professional fishing guides around the lake. Out fishing one morning in the fog, I heard somebody say, 'Two women are lost.' I could well understand what that meant. The lake is gigantic–an estimated 184,000 acres. Getting lost in such an area could be disastrous in any kind of weather, but moreso in heavy fog. I asked for any location information and a guide told me he was near the area, but was having to get his boat out of the water because it had hit a log.

"Fortunately, I was near the area and motored toward that point. I stopped my outboard and listened. I heard some hollering and quickly went over. I found the ladies and asked if they were the ones lost. 'You're the best looking man we've ever seen,' they replied. I towed them to the nearest marina. They, of course, wanted to pay me and all that business, but I told them I didn't answer the call for money. I was just glad I happened to have a CB."

Leisure time has come to be both a "curse" and "lifesaver" to all humans. Shorter work weeks give more time for leisure activities, but what to do with the time is something else. Resort areas are taxed to capacity almost every weekend with pleasure seekers wanting to get away from it all.

More and more churches are finding that there is a need to go out where the people are. Let's face it–more than likely they will not be leaving the resort to go to church! So, in many parts of the country, churches are holding services in resort areas themselves. Now here's where the CB comes in. As the tourists are moving into the areas on Fridays and Saturdays, church members with CB's greet the campers (over the air) and tell about the service for Sunday, giving the location, time, and speaker. The "Lake Missionary" at Toledo Bend says it works!

10. Being a Good Samaritan

A doctor friend in Pineville relates an experience of a friend in Kentucky. The friend was an avid CB'er. Leaving a friend's house one snowy evening he began modulating with a person on a base station. Suddenly, all contact was lost—without apparent reason. The base station operator immediately contacted friends and organized a search party.

They found that there had been an accident. A companion of the lost person was killed, the car having been side-swiped. The driver (the one who had been talking to the base station) was thrown from the vehicle into the snow. He was found within twenty minutes of the loss of contact. A doctor later said that if the driver hadn't been found when he was, he would have died from over-exposure within thirty minutes. Although not related directly to witnessing, this account does illustrate the ministry of men caring for their fellow human beings.

The rescuers were lauded but refused to accept all the credit. They said they simply did what any other person would do. I would like to think that is true, but in too many cases, it isn't. CB'ers have a camaraderie that seems to be unique, with friendliness being the most cherished ingredient in their fellowship. The willingness to help is so evident. Perhaps this kinship can be explained by a poster I once saw. It said

It's Nice to Be Important
But More
Important to Be Nice!

Another incident illustrating a positive use of CB is shared by Waldo Woodcock (Georgia New Day), Church Training secretary for the Georgia Baptist Convention.

"A teenager was hurt in a motorcycle accident and received a broken leg. His father needed to be notified, but he was fishing on a lake forty miles from home. The boat was equipped with a CB, and a base station was trying to make contact. Hearing the 10-77 (negative contact), I offered to 10-5 (relay) the message since I was heading in the direction of the lake. I was successful in reaching the father in ten-fifteen minutes, and in a short time he was on his way to the hospital. He received reports all along the way."

In all my conversations and modulations over the airways, I have yet to come on a situation where a fellow CB operator refused to offer assistance when assistance was needed. I'm sure there have been instances of meanness and disregard for fellowmen, but my experience has been just the opposite.

Consider the story of the "Good Samaritan"

Luke 10:30-37

"Jesus replied with an illustration: 'A Jew going on a trip from Jerusalem to Jericho was attacked by bandits. They stripped him of his clothes and money and beat him up and left him lying half dead beside the road.

'By chance a Jewish priest came along; and when he saw the man lying there, he crossed to the other side of the road and passed him by. A Jewish Temple assistant walked over and looked at him lying there, but then went on.

'But a despised Samaritan came along, and when he saw him, he felt deep pity. Kneeling beside him the Samaritan soothed his wounds with medicine and bandaged them. Then he put the man on his donkey and walked along beside him till they came to an inn, where he nursed him through the night. The next day he handed the innkeeper two twenty-dollar bills and told him to take care of the man. 'If his bill runs higher than that,' he said, 'I'll pay the difference the next time I am here.'

"Now which of these three would you say was a neighbor to the bandits' victim?" The man replied, 'The one who showed him some pity' " (TLB).

Let me paraphrase this Scripture reference and see if it has possible connotations for the Christian CB user.

A salesman was going on a trip from Shreveport to Dallas, and he had car trouble on Interstate 20. But he was lucky because he had a CB radio. He got on the unit to call for help. The first unit he saw with "ears" never did even respond to his "break, break" for assistance. In fact, the driver got over in the other lane to avoid getting too close.

The second one, a 4-wheeler with ears, came back "wall-to-wall and tree-top tall," but his only response was that he was late for an appointment and that he would try to 10-5 the message to someone else.

But a third one came along, an 18-wheeler, and hearing the 10-34 said he would be glad to help. Upon finding out the trouble, he went to the next town and found a mechanic. The 18-wheeler brought the mechanic back to the disabled car, stayed there till it was fixed, offered additional assistance if needed, then went on his way.

Now, which of the three would you say was a good neighbor to the CB'er needing assistance? A reply came: "The one who showed him concern." Then Jesus said, "Go and as you go, help when you can."

On numerous occasions, I have helped fellow CB operators (and non-operators) by coming to their assistance. I vividly recall one. A young lady, who really didn't know much about talking on a CB, called for help. I was on I-20 east of Shreveport and could tell by the 10-2 (strong signal) that she was close by. I came back and asked her 10-20 and type of vehicle. She said she was on I-20 near Shreveport and was in a yellow Volkswagen.

I pulled up behind her, asked what the trouble was, and found out she had run out of gas. I said I would be happy to go get some for her and she was most appreciative. The next exit was only about two miles west, so I took off and said I would be back as soon as I found a station. Just minutes after leaving her, I heard her calling again for assistance, just as if I had not stopped. I couldn't quite figure it out, so I contacted her again and reassured her that I would be back–with the gasoline.

Some ten minutes later, I pulled up behind her. After putting the gas in, she apologized for not believing I would come back. Of course, she tried to pay me, but I told her I didn't stop to make money. I simply said, "Aren't you glad you had a CB?" She replied, "Yes, I've fussed so much at my husband for playing with that silly box, but never again. This is one time I'm glad it was handy." A good motto for Christian CB'ers might read like this:

YOU DON'T HAVE TO BE A CHRISTIAN TO BE HELPFUL,
BUT A HELPFUL CHRISTIAN YOU SHOULD BE!

The "Samaritan" did not carry the label of "Christian" in the story, nor did the "18-wheeler" in the paraphrased version. But their actions were Christian, and that's the difference. The Samaritan didn't expect payment for services rendered, nor did the 18-wheeler. Both responded out of a sincere desire to help.

Christians are supposed to be good citizens, willing to help others in need. We do not necessarily have to talk to someone about their relationship to Christ before we help them (though we should if led by the Holy Spirit). We can merely help them by getting them gasoline, a mechanic, or directions. In so doing we are ministering to our fellowman.

Unfortunately, use of the CB is sometimes abused by those who have little regard for others. Illegal poaching of wild game and other corrupt practices have been reported. I recently read in the newspaper that some CB'ers were blocking lines of communication on several channels when persons were trying to get help for a heart attack victim. They probably thought it was cute and were oblivious of the consequences of their actions.

Such abuses have been known to draw violent reactions. A recent newspaper account read:

FIGHT OVER CB CHANNEL USE LEADS TO PERMANENT 10-7

"Blue Rover" is charged with the first-degree murder of "Bear Tracks." It seems that the two CB'ers had a running feud over the use of a particular channel. "Blue Rover" is accused of deliberately running down "Bear Tracks" with a pickup truck in a parking lot. The "Blonde Bomber," an eye witness, gave this account: "The two men had a heated argument over the CB and decided to meet at a certain place. They exchanged punches before 'Blue Rover' jumped in his truck and ran over 'Bear Tracks' two times. He died three days later." Far better it would have been had they heeded Colossians 3:8 and "put off all these things: anger, wrath, malice, blasphemy, (and) filthy communication."

Another incident in September of 1975 dramatizes the misuse of the CB. The scene took place in Louisville, Kentucky, and the setting was antibusing demonstrations, which eventually led to a number of riots.

The county law enforcement officials, state and city police, and National Guard were called to the scene to help keep order. Soon, it was evident that the rioters and demonstrators knew every move made by the police officers. The Louisville police chief discovered that the rioters were using CB radios to evade the law and organize the demonstrations.

Anti-CB orders were given to jam the channels being utilized by the protestors. CB communications were monitored and jammed when necessary to momentarily disrupt the movement of the riot. One of the communications monitored was between riot leaders and a truck driver, coming into the city with a load of ammunition.

These violent, aggressive abuses were committed by people who intentionally set out to do harm. Sometimes we look at individuals like this and point our fingers without realizing that we, too, sometimes fall into the category of "law breaker."

Consider a young man who was a member of a Bible Institute who was traveling on a highway at a terrific rate of speed. Overtaken by a traffic cop, the student was ordered to curb his car. Timely and golden were the words spoken by the officer to the student. On the student's vehicle were printed the words, JESUS SAVES. Said the officer, 'Your willful violation of the law; your endangering innocent lives, as well as your own, by the illegal and dangerous speed at which you were traveling, is greatly in variance with the words printed on your car, and brings no honor to the One of whom your motto speaks?' "[10]

When I first began using my CB, I jumped on the band wagon of helping to spot the "smokeys," avoid them, and help others avoid them. Even when I didn't go over the speed limit, I still listened for "smokey reports" just for the game of it. It was fun–until I came face to face with Romans 13. It happened like this.

I had been asked to supply preach at a church in north Louisiana. I had decided that a message on patriotism would be in order because of the Bicentennial. And I began to study. My research led me to Romans, where I found:

> "Obey the government, for God is the one who has put it there. . . . So those who refuse to obey the laws of the land are refusing to obey God and punishment will follow. For the policeman does not frighten people who are doing right; but those who are doing evil will always fear him. . . . The policeman is sent by God to help you. But if you are doing something wrong, of course you should be afraid, for he will have you punished. He is sent by God for that very purpose. Obey the laws, then, for two reasons: First, to keep from being punished, and second, just because you know you should" (Rom. 13:1-5, TLB).

If the policemen are God's ministers and are working to help people, then I'm working against God if I break the law. When I speed on the highway, I not only break the law, but in telling other law breakers where the "smokeys" are, I am thwarting justice.

When I preached that sermon, at least one person made a decision–me! Since that decision, I have not used my CB to hinder the work of God's ministers (policemen). Public service messages are now being shown on television with these thoughts.

The 55-mile-per-hour speed limit has saved thousands of lives. It has saved millions of barrels of oil, but the one fact people too often forget is–the 55-mile-per-hour speed limit is the LAW.

Hopefully, positive use of a CB will be established in this book (witnessing, ministering, assisting, etc.). The negative aspects will be pointed out so that you may decide on your own the best use for your CB.

Bob Waldon (Bob-a-long), Director of B.Hi.U. work for the Louisiana Baptist Convention, shared this conversation. It further supports the idea of God's ministers found in Romans 13.

"I was talking with a highway patrolman recently and asked him what he thought of CB radios. The patrolman said he thought it could be the best thing that has happened to car travel in quite some time, were it not for the misuses that are occurring. He went on to mention the fact that when a 'smokey' sets up a speed trap or a road block, 18-wheelers and 4-wheelers often use the CB to tell each other about the location. Therefore, the very ones who are breaking the law slip by. Most persons who are quick to give 'speed trap' locations think they are being of service, but this is not the case. This is CB abuse."

The patrolman's next statement offers some food for thought:

"Would you want a drunk driver on the highway driving 90 and 100 miles-an-hour finding out 'smokey reports'? This type of driver doesn't need to be on the highway, but seldom gets caught. Suppose your daughter, son, husband, or wife is killed by a person guilty of breaking the law in this manner? Wherein lies the blame? . . . with the drunk driver, or the well-meaning CB'ers spotting 'smokeys' for him. It's something to think about–often!"

If we claim to be Christian, then we should be found guilty of "acting like a Christian." Owning a CB radio requires a license, but does not validate a "license to break the law."

Peter sums up the role of the Christian in life and government with these remarks in his first epistle:

> "You are free from the law, but that doesn't mean you are free to do wrong. Live as those who are free to do only God's will at all times. Show respect for everyone. Love Christians everywhere. Fear God and honor the government" (1 Pet. 2:16-17, TLB).

How many times have you heard the statement? "The good kids never get any credit for what they do, but let one step out of line, and it's on every front page in the country." It seems we hear of such reviling circumstances all too frequently. These are the ones that sell papers and make good television news. Worthwhile endeavors like the Mississippi Coast REACT often go unnoticed. REACT is a volunteer Citizens Band team which monitors the Federal Communication Commission's official emergency Channel 9. The Mississippi team voluntarily works twenty-four hours a day, offering assistance, traffic information, road conditions and directions.

When a REACT member gets an emergency call, he quickly contacts local police or the highway patrol by telephone and gives the necessary data so that assistance can be provided.

The Cenla (Central Louisiana) REACT operates on a similar basis, affiliated with the REACT headquarters in Chicago. I recently had the opportunity to meet a man who offers a very fine public service through the Cenla REACT. His story follows in the next chapter.

Bill Keating has not let poor health keep him from being a service to the public. For several years he has used his CB to assist motorists throughout the Central Louisiana area.

11. Public Service

"A Godsend to shut-ins!" Mrs. Bill Keating used those exact words to express what CB had done for her husband. "After an active business life, my husband has been in this house for three years now with heart disease and emphysema. He was also very active in the Methodist Church, having taught a men's Bible class for years before yielding to poor health."

Bill Keating has a CB right by his bed. His son gave it to him two years after his confinement. "I had watched television and listened to the radio for about as long as I could stand it," he said. "My son told me I would really enjoy the CB and how right he was. Now, I have something useful to do with my time, and I'm like a kid with a new toy."

Mr. Keating, better known as the "Wheelchair," has received numerous awards for service to the community of Alexandria (Louisiana), as he is active in REACT and the promotion of good will over the CB. "I work nine hours a week every third week on the REACT channel, and the work is really rewarding. I get a certain satisfaction out of helping folks in need and reporting emergencies that help to make our area a safer and better place to live."

When Mr. Keating is not on REACT duty, he is at work monitoring the popular channels to lend assistance, give directions, or just talk with the "ratchet jawers." He continued, "I keep an up-to-date city map right handy to give directions on street locations. I also keep the telephone numbers of police and fire stations close by in the event of emergencies."

Your area may not have a REACT team, but you could be the one to start it. Contact the FCC office nearest you for information. It's not an organization of the church, but its services are reflective of Christian ideals. The following report gives evidence of this fact.

In *The Mississippi Press,* it was reported that CB clubs in almost every city you can think of are raising money for good local causes. In San Diego, CB'ers joined 7-Eleven Stores' promotion to raise $6,000 for muscular dystrophy patients. The drives are relatively small in other cities, but are characterized by dedication to individuals that really need help, gathering $100, $200, or even $10,000 for a cancer patient, eye operation, or certain widow.

The *Press* further stated that the efforts are not organized like Scouting paper drives or Heart Fund campaigns, but are spontaneous in nature. Most of the appeals are made over CB radios or at monthly meetings, and the results are really unbelievable.

Good Samaritans are not hard to locate. But reports of their activities may be, for the deeds are "behind the scenes." Come to think of it, that's not all bad!

12. Witnessing

When I think of witnessing over a CB radio, my thoughts immediately turn to the word "life-style." For witnessing is not a set of rules—hard and fast—that are memorized so you can quickly catch a prospect off guard and let him have the "full load." But rather it is a combination of all you are and hope to be in Christ. Yes, it is on occasion a retelling of your own personal experience with Christ. Yes, it is on occasion a quoting of Scriptures and yes, a sharing of how Christ is working in your life today. It's even more than all of these, but there are no set rules or times when you use one or another. YOU are the *witness. You,* guided by the Holy Spirit, will be the one to determine when and how to verbalize your witness. Your "life-style" as a Christian will help determine situations in which you find yourself. The ideal is to be able to be a positive witness in whatever situation that arises.

Lew Reynolds, director of Sunday School work in Indiana, shares the following testimony of two situations in which he found himself.

"On a recent Sunday morning drive to a supply preaching assignment, my handle (Good News) caught the attention of an out-of-state CB'er, and he questioned my line of work. I shared briefly my opportunity of denominational service and the fact that I was on my way to a preaching engagement. He, along with his family, shared with me and others on the channel their personal experience with Christ. Two other families joined the convoy modulation, tying up the channel for miles in a free and easy testimony service with each telling me one at a time that they would be praying as I preached, and I in turn would pray for them (by handles) as they traveled.

"One of the most enjoyable parts of having a CB comes when you get a chance to modulate with another Christian concerning your faith in Christ. You never know who is listening, but when you share of the riches of Christ, it is promised in God's Word that 'the Word of God will not return void.' "

Lew Reynolds continues his testimony with another account.

"Probably the most dramatic experience came this spring on 65 South from Chicago. It was one of those nights when no one was on the channel. In fact, it was downright spooky because there was so little traffic on this major north-south expressway. Add to this a thunderstorm off into the west with long streaks of lightning.

"Then over my left shoulder (East) I suddenly was aware of one of the most beautiful spectacles in the sky that I have ever witnessed. I pulled my car over on the shoulder and became awed at a sight I had never seen before. I hit my CB to see if anyone north or south was seeing the sight with me. The first man said it was the Northern Lights and gave some scientific information concerning this rare phenomenon. I responded with a comment of how the good Lord could put on a better fireworks display than we could ever see on the Fourth of July.

"Another said that is nothing like it is going to be when He comes again. Someone else started singing, 'Mine eyes have seen the glory of the coming of the Lord.'

"Someone else was saying 'Praise the Lord! Praise the Lord!' over and over again. Someone else was reading or quoting from Revelation–some of the things John saw in his vision. I felt a strange togetherness with people I did not know, nor could I even see the people I was talking to. The spectacle in God's great sky, plus the witness of his people and the CB channel made this a night I shall not soon forget."

Please notice that Reynolds did not come on too strong and in fact, didn't even initiate the conversation on the first testimony. His witness was natural and positive in every way.

There are basically four prerequisites for witnessing whether it's over a CB radio or face to face with someone.

1. *You Need to Be an Eyewitness to the Fact.*

The apostle John wrote the following:

"Christ was alive when the world began, yet I myself have seen him with my own eyes and listened to him speak. I have touched him with my own hands. He is God's message of life. This one who is life from God has been shown to us and we guarantee that we have seen him; I am speaking of Christ, who is eternal life. He was with the Father and then was shown to us. Again I say, we are telling you about what we ourselves have actually seen and heard, so that you may share the fellowship and the joys we have with the Father and with Jesus Christ his Son" (1 John 1:1-3, TLB).

In his book, *Christian Doctrine,* W. T. Conner makes the statement, "A knowledge of the truth is necessary to propagate the gospel." He goes on to say, "One of the fundamental impulses of the Christian life is the impulse to propagate the gospel of Christ. It was pointed out that becoming a Christian was a rational and voluntary act based on a knowledge of the gospel. The one propagating the gospel then must be able to give to the one whom he would bring into the Christian life an intelligent conception of what it means to be a Christian. The propagator of the gospel; therefore, must have an intelligent statement of the same."[3]

Dr. Conner is simply stating the fact that if a person is going to share Christ, however he does it (in the pulpit, in a class, in business, or over a CB radio),then that person needs to be well-grounded and based in the faith of Jesus Christ. He further implies that the presentation of the gospel based on what the Holy Spirit does with that presentation makes all the difference.

There is no way we can share with people what we do not have ourselves. It is virtually impossible to be that good an actor or impersonator, for we must first experience new life in Christ and strive daily to "walk with him." Then we have a message or witness to which we can attest. We too can then make the statement with John: "I guarantee that I have seen Him—the one who gives eternal life, for He abides within me."

And when he abides within, the word life-style again comes up. Dr. David Haney of the Brotherhood Commission gives the following definition: "Lifestyle Evangelism is a new approach to our unending assignment as Christians. It says that evangelism is not just something you 'go and do;' but rather it is something you 'do as you go.' It is the kind of evangelism which takes place in the normal flow of one's daily events." [18]

2. *Be Sold on the Importance of Your Testimony*

The apostle Paul's use of his testimony in the book of Acts is impressive. Almost every time he spoke, he used his own personal witness of what had happened to him. Your testimony is just as important to your life-style of witnessing as Paul's was to him.

It doesn't have to be earth-shaking, but it does

have to be *you.* And since it is you and happened to you, it will command attention. Your unique experience also will not have to be defended, because it happened through a personal experience with God.

Bob Lawrence (Colorado Tune-heister), director of the Church Music Department in the Colorado Baptist Convention, gives the following account.

"We had a young man from Georgia, who was a pastor in Montana. He got interested in CB and put a base in his home and a mobile in his pickup. He got to talking with the 18-wheelers going through on Interstate 90 and began inviting them to drop by his house for coffee. He was fortunate in living on a street which was the width of two streets and had plenty of room for them to park in the middle, leaving passage on both sides.

"One man came by and learned that he could trust Wayne and then you know what happened? Before long, when a fellow had difficulty, he would call the 'Rev.' and would come by and talk with Wayne. He has had as many as five 18-wheelers parked in front of his house at once with the guys sitting around his dining room table drinking coffee.

"He was able to win one of them to the Lord, and this man quit trucking and is now a member of the church in Butte, Montana. I understand he is making a very fine Christian. He wants to try to minister to truckers, but because it is such a difficult thing to do, he has not gone too far with it.

"I remember Wayne telling about a fellow who lost his little girl in an accident. Coming through town, he got on his CB and said, 'Rev., I need to come by and talk with you real bad this time because on the last trip home, we buried our little girl, and I sure do need to visit with you.' "

Wayne's testimony proved to be very important. So is yours! Share his love–day by day!

3. *Don't Turn Off Your Listener with Words He Doesn't Understand.*

Such words as "salvation," "saved," "born again," "walked the aisle," or "trusted Christ" are not understood by non-Christians. Don't turn them off from the very beginning. At this point, let me throw in a negative testimony, but only to make a positive point.

"Since I do not own or use a CB, it is possible that my reaction to your request for information on witnessing over a CB is way off base. I get very annoyed at Mormons, Jehovah's Witnesses, Moonies, and others because of the frequent high-pressure pestering we are subjected to at home and in shopping center parking lots. I don't want to be guilty of doing the same thing, nor act in such a way that the public would have reason to lump us in groups like this. For me, it is cheap!"

Needless to say, the person making the above statements will probably never get into CB, but his statements reveal an important caution. The caution has to do with the statement about "high-pressure pestering."

One can never get on a CB radio and try to overpower someone by bluntly communicating to them, "I want you to become a Christian," or "I want you to become a believer in my God." The true witness comes when the person is carrying on a conversation using the usual CB jargon. Then, as the Holy Spirit directs and as the opportunity comes, one can talk about his own relationship to Christ, the church, his work, and how the Lord has blessed him. As you examine further the many testimonies in this book, perhaps you will begin to see how opportunities to share Christ do arise.

The high pressure method won't work and should never be used. You think a person can turn you off when you're witnessing to him face to face, but not as easily as a person with a CB. They can either turn off their unit, switch channels, stop the conversation by not saying another word, or make the statement, "Someone walked on you." Witnessing over a CB and/or ministering in the name of Christ must be done discreetly, and out of love for fellowman, not because of what you hope to gain, but because of what Christ has done for you.

In response to my request on information about witnessing over CB, many made statements that Christians should act like Christians, be brief and courteous. In Titus 2:8, we find this statement that certainly backs up the above. "Your conversation should be so sensible and logical that anyone who wants to argue will be ashamed of themselves because there won't be anything to criticize in anything you say!" (TLB).

Dr. Leonard Sanderson, director of the Office of Evangelism, Louisiana Baptist Convention, has had some memorable experiences with his CB radio. When asked about his handle, "Rollin' Chaplain," he said, "When people find out that they are modulating with the Rollin' Chaplain, one of two things will happen: either their tongue will come loose at both ends, or they will clam up all together. Many will ask about my ministry or bring up biblical topics—many of which lead to very lengthy and creative discussions. Others will sign off very quickly with: It's good to make contact, Rollin' Chaplain—have a good day and a better one tomorrow and go 10-7."

Dr. Sanderson also says that a few times he had seen disorders on the highway and felt like they ought to be reported. "One man was trying to chase another off the highway and almost succeeded. Several times, he almost caused a wreck. I made contact with others and through a 10-5 was able to summon some help. This incident alone convinced me of the need of having a CB."

Dr. Sanderson feels that his handle has been a major asset in reaching people for Christ as it opened up conversations in rest areas and camping spots. "I get a chance to talk to people face to face, many of whom I had talked to earlier in the day over my CB," he says. "This was one of the reasons why I wrote a CB tract. I wanted to put something tangible in the hands of people that would be of some significance."

The tract of which Dr. Sanderson speaks has been distributed all over the country and has been utilized a great deal in Christian CB clubs. He emphasized that any CB operator should welcome opportunities to share this tract or other tracts to proclaim God's love to all men.

The words in the tract are not confusing, nor are they irrelevant. The Scripture references are clearly and cleverly tied in with the 10 signals. You may never have the opportunity to give your testimony over the airways, but you can leave a tract with a fellow CB'er. *Dr. Sanderson's tract is available at printing and mailing costs from the Louisiana Baptist Convention, Box 311, Alexandria, Louisiana 71301.*

4. *Know the Power Source*

In Acts 4:13-18, we find Peter and John as they boldly preached the gospel. They were being chastised for their efforts, and in verse 18 it says, "And they called them and commanded them not to speak at all nor teach in the name of Jesus."

Many Christians somehow feel that this verse is intended for them because they never speak out for Jesus Christ. This is against the New Testament teachings of sharing your faith. So, if an opportunity arises for you to share what Christ is doing in your life, you should follow the example of Peter and John by remembering our "power source" in Christ and speaking with unashamed boldness—over the CB radio and in all situations.

> "As the rain and snow come down from heaven and stay upon the ground to water the earth, and cause the grain to grow and to produce seed for the farmer and bread for the hungry, so also is my Word. I send it out and it always produces fruit. It shall accomplish all I want it to, and prosper everywhere I send it" (Isa. 55:10-11, TLB).

We are admonished time and time again in the Scriptures that we are not to be found *successful* but *faithful* in our witness for Christ. God will take care of the success in his own time and season.

Jesus prefaced the Great Commission with these words: "All power is given unto me in heaven and in earth" (Matt. 28:18). If we go with the gospel, He will supply the power source. Such sources of power are the Bible, prayer, and the Holy Spirit working in and through the lives of the persons to whom we speak. There is no way we can convict a person through threat or manipulation. The Holy Spirit will accomplish all he wants if we present the gospel.

Richard Sims (The Counselor) reported these experiences of sharing Christ:

"I met another CB'er by the name of 'Ticket Taker' on the wire. Ticket Taker indicated that he owned several theatres in Dallas, north Texas,

Oklahoma, and the Kansas area. We continued our conversation on the highway for about 300 miles, and then he asked if I would like a cup of coffee. We stopped, and over the coffee cup I had opportunity to ask him about his relationship to God and also the pictures he recommended for use in his theatres.

"I don't know if I did a lot of good except that he was intensely interested in my ministry. He shared some problems that he was having with his family because of his being on the road. I believe I helped at this point, for I placed him in contact with a local minister, and Ticket Taker promised me that he would contact him.

"Another was a trucker who had just lost his mother and was asking about directions. He told me that he was going to stop in his home town long enough for the services, and it was an opportunity for me to minister to him.

"Our conversation went like this: 'The Lord has promised that he would take care of his children, and losses are easier to take if you have the Lord as your Savior. I continued on in 1 Thessalonians and of course, the 23rd Psalm.' "

Richard says that if he had to make a choice between a CB and a spare tire, he would choose the CB! Perhaps no one will ever know what happened to the "Ticket Taker" or the one who lost a loved one. But God's Word will not return unto him void.

13. What Do You Say? How Long Do You Talk?

There are no "pat" answers to these questions. As you examine the various testimonies in this book, I hope you analyze them by asking some of the following questions:

(1) Who started the conversation?

(2) Was the person doing the witnessing "pushy"?

(3) Was the conversation courteous or tactless?

(4) Was it too lengthy–would it have been more effective to terminate earlier?

(5) Who did most of the talking?

Use the above questions to analyze another testimony of Lew Reynolds (Good News).

"On another occasion, late at night coming out of Chicago, a convoy of four or five vehicles enjoyed getting acquainted, kidding back and forth, and just enjoying the ride down the strip. One of the men who went by the handle 'Sergeant Bear' was coming in wall to wall and I expressed appreciation and complimented him for clear, deep, professional-like diction. He told us he was a sargeant with the state police on his way to West Virginia to attend his mother's funeral.

"This gave me an opportunity to express my concern and to share some Scripture verses with him, including Psalm 46, from memory. He in turn expressed appreciation and gave his own testimony of what the Lord meant to him—in his work and from day to day.

"This brought in at least four other vehicles including 'Lamplighter,' an 18-wheeler, asking for all four of us to pray for him. He was in the process of a divorce and acknowledged that much of the trouble could be blamed on him. 'Sergeant Bear' offered to stop for coffee, but the 'Lamplighter' indicated that his schedule would not permit."

(1) *Who started the conversation?* "Good News" began the witnessing part by expressing Christian concern, but only after "Sergeant Bear" volunteered the information about the death of his mother.

(2) *Was it pushy?* Because "Sergeant Bear" expressed appreciation and gave his own testimony, I don't feel that it was. Being perceptive to the

(3) *Was the conversation courteous or tactless?* Both "Sergeant Bear" and "Good News" were

(3) *Was the conversation courteous or tactless?* Both "Sergeant Bear" and "Good News" were courteous in every way. Just suppose "Lamplighter" made this statement at the end of transmission, "Well, you may not talk to me, but you better get right with God, or you're going to have

even bigger problems." This was not the case. The modulation ended on a positive note, leaving the door open for someone else down the road who might have a chance to pick up where "Good News" and "Sergeant Bear" left off.

(4) *Was it too lengthy?* No, for the same reasons as outlined in question (3).

(5) *Who did most of the talking?* "Good News" did the most, but the content was good and commanded attention.

Here's another one for you to analyze. It's given by Guy Ward (Joe Baptist), secretary of the Religious Education Department in the Utah-Idaho Baptist Convention.

"Traveling to Glorieta, New Mexico, to attend a conference, I came upon a motorist with the inscription 'Do You Know Jesus?' in large letters on his rear bumper. I contacted him to say that I liked what he had written on his bumper.

"Through the ensuing conversation, we were able to share with all CB'ers within earshot about Jesus and his wonderful gospel. Yes, I have found that CB can be used in witnessing. I wouldn't be without one!"

This testimony was not long at all, but a witness was given on the part of both operators. The benefactors hopefully were those who were listening. The story doesn't indicate whether or not anyone was listening, but from my experience, I'm sure there were some "ears up." This took place on a highway, and most operators will monitor Channel 19 on the road. Listening is part of the game, and it's also a significant thing to remember when you do have opportunity to share your faith in some way. You never know who is listening or the problems they are carrying. A personal friend of mine, who does a lot of counseling, told me this. "You wouldn't believe the *garbage* people carry around with them (guilt, failure, sin, etc.). A crying need of these people is a listening ear–someone to empathize with them and perhaps offer a helping hand. A lot of these people have CB units. They may never be reached or helped through any other means. Don't pass up an opportunity."

14. Spot Announcements

Some of the characteristics of CB operators are their "sign on" and "sign off" messages. This is especially true of the 18-wheelers. You'll hear such comments as:

"Top of the day to you, the one 'Cajun Cowboy' is now on the channel."

". . . 3's, 8's, and all them good numbers to you."

"Keep it between the ditches, this is the one 'High Pockets' doin' it to South–bye, bye."

"May the good Lord take a likin' to ya, the 'Blue Bird' southbound and down."

". . . have a good day and a better one tomorrow. The 'Yellow Jacket' truckin' North with the hammer down."

For some reason, ears are glued to the unit when jargon like that starts. It catches your fancy! A lot of the talk doesn't make much sense, and it may be hard to understand, but I find myself listening.

Short greetings or salutations with Christian overtones can be effective tools for sharing the Good News. The "Lake Missionary" signs off with, "May God bless you," and I have often said farewell with, "The Lord's blessings on you as you travel." Others that could be used are:

"Let the Son shine in–God's Son that is!"

"This is 'Child of The King' on the road to Glory." The Holy Spirit can use little bits of conversation like this which may pave the way for a broader witness on down the road.

Illustrations worked into the course of a conversation can get across a point in a short amount of time. The time has to be right and the listener receptive. For example, an analogy can be made between parts of a CB unit and Christianity. Here is a possible conversation model between two "good buddies."

"Breaker one-nine (19) for that 'Jubilee Kid.' This is 'Frogman' listening."

"Go ahead, 'Frogman,' greetings to you."

"Going great, 'Jubilee.' Just thought I'd give you a shout, shout. Heard you modulating back up the road."

"Yeah, that's a 10-4. You're coming in wall-to-wall and tree-top tall–whatcha' shooting with?"

"Got me one of the 723's, 'Mr. Jubilee,' with a whip."

"10-4, Frogman, those aerials are real important to good reception. Reminds me of something."

"What's that, 'Jubilee'?

"Just as we need aerials for our units to operate properly, we also need aerials in our lives to communicate with God. The strength of the communication depends on how well we're keeping our 'ears up.' "

"A big 10-4, 'Jubilee.' That needs to have a 10-5."

This conversation, to my knowledge, has never taken place, but it could. Notice the design and direction as the modulation progressed. "Frogman" did the contacting initially, wanting to talk with the "Jubilee Kid," whom he had heard back up the road. This will happen quite frequently as you travel. You get within earshot of a conversation and after catching the "handles," break for one of those units. "Jubilee Kid" was ready and willing just to pass the time of day with small talk, but he also had some ideas on making the modulation more meaningful.

The term, "whatcha' shootin' with?" is heard often and is the CB'ers' way of asking what kind of radio are you using. When "frogman" gave the type and some information on the aerial (a whip is a long bumper-mount type antenna, "Jubilee Kid" felt like he could go into the analogy.

In most of the testimonies found in this book, the conversations usually moved toward a meaningful witness, but only after friendship had been established. The surfacing of a key word or question may offer the opportunity to share a "spot message."

The experience that I had with "Laredo," mentioned in the Introduction, falls into the category of friendship. The key statement concerned his son being in the hospital in New Orleans. I have thought through the course of that conversation many times, and one question comes up each time. "What would I have said if I had been talking with

'Laredo' face to face?" I'm sure I would have told him that I would pray for his son, but I don't think I would have prayed with him, unless it would have been in a Sunday School class or other conducive situation. Perhaps the fact of not being seen, only heard, offers some security. I never saw "Laredo" or his wife, but I did see a development of a trusting relationship, if only for a short time.

Calvin Cantrell's testimony in the beginning of this section also supports the theory of security. The blind base station operator felt secure behind that CB mike, for the blindness was not known, except to persons who knew him personally. Calvin even made the statement that, "If I had known he was blind, I would have treated him differently–pitied him just like others were doing." The blind man opened up with Calvin, and the result was an eternal salvation experience.

15. Christian CB Clubs

Christian CB clubs are springing up all around the country. One such club has recently been started in Casper, Wyoming, by the four Southern Baptist Churches there. Herbert Whitten, pastor of the Mountain View Baptist Church, shared the following with me.

"Our main idea in beginning the club was to improve the image of CB operators in the Casper area. Most people saw all CB'ers in our town as hobbyists whose only pastime was partying. The four pastors decided to see if there was any interest in beginning a Christian CB club–and there was! We meet each month at one of the churches to discuss various projects, most having to do with ways we can better serve the needs of the community with a Christian spirit.

"We monitor one channel specifically for contact and assistance. Counseling points are designated when referrals are made. The acceptance has been most gratifying."

Seeing this as an opportunity for many cities and areas, I asked Mr. Whitten if he would share the bylaws and guidelines established by the club. He was most happy to share, hoping that they might help other clubs get off the ground. These guidelines may be found in the Appendix at the end of this book.

Some activities that might be included in the list of possibilities for Christian CB clubs are:

(1) HOLIDAY TRAVEL SERVICE–CB operators traveling through the city on various holidays would be greeted and invited to stop for coffee and doughnuts at a convenient roadside area. There, friendly hospitality would be given and tracts handed to the travelers. The purpose of the activity is two-fold.

A. Travelers need to rest for safety reasons.
B. Witnessing opportunities may surface.

(2) RENDER PUBLIC SERVICE–Join in community projects like fund-raising and anti-litter campaigns.

(3) EMERGENCY WORK–Become involved in REACT, a nationwide organization that monitors Channel 9, the emergency channel. Christian CB'ers need to be involved in activities of this nature in order to exercise good citizenship.

I hope the probability of your using CB for ministry and witnessing has been increased as you have read the previous sections of this book. You may already have several experiences that you could share, and to this I say, "Praise the Lord." But if you've been skeptical or haven't had a chance, don't fret. If you're willing to be used of God to share Christ, the opportunity will come—then it's up to you.

PART THREE

"10-33"—Help Me Quick!
(Everything You've Always Wanted to Know About CB Radio Without Going into a Lot of Detail)

CB is no longer a "fad"—it's here to stay. A projected three billion dollars in sales for 1977 is reason enough to say that it's past the fad stage. Count the number of CB aerials in a shopping center or church parking lot. Have your kids get into a game of counting antennas the next time you're on a long trip, and you'll be amazed at the number.

The craze has taken over the country and become what some have said is a "nationwide party-line." You can't pick up a newspaper, magazine, or look at your television set very long without being bombarded by an advertisement about CB radios. But with all of the advertising and promotions that are going on, some are still not quite sure about this latest electronic "gadget." I believe that this part will be of help to anyone interested in CB, but it will especially help the beginners and those who think they may want to invest in a CB unit.

16. So You Want a CB?

Have you ever wondered how it all got started? Contrary to what you may have heard, it was not an overnight sensation! It all started back around 1945 when E. K. Jett invented the first CB. The FCC assigned frequencies for Citizens Band use, but no one was really interested. The reasons are still not clear, but between 1945 and 1958, very few licenses were issued. Of course, very few radios were sold. In 1958, things started to pick up a little when "Uncle Charlie" (FCC) established the Class D Citizens Band. It allowed a low-cost, easily available, short-distance communication system for business and necessary personal use. However, between 1958 and 1973, still only about a million licenses were processed.

A number of events triggered the boom in sales which began in 1974. First of all, the Arab oil embargo hit the United States like "Sherman going through Georgia." Lines at the gas pumps began to form, and people started looking for ways to find out where the short lines were—and even where the gasoline could be found. About that time, too, the 18-wheelers were beginning to be hard-pressed for "motion lotion" (diesel fuel) which directly affected their livelihood.

Second, the "double nickle" syndrome (55-miles-per-hour speed limit) hit nationwide. The truckers said they couldn't make a living driving that slow—too much time was involved. Individuals simply didn't want to slow down. So a combination of these two factors brought the CB radio into the limelight. Almost overnight, the pendulum swung from almost no one having CB's or knowing about them to John Chancellor's announcing to the whole country about the truckers' strike—a strike promoted and coordinated through a gadget called a Citizens Band Radio. The strike was in protest of outrageous fuel prices and the 55-miles-per-hour speed limit. The strike neither lowered the price nor increased the speed limit, but it did cause CB to become a household word. I'm reminded of a statement that I heard recently, which helped to illustrate the sudden popularity of CB:

> "Yesterday I didn't even know what a CB operator was, but today I are one . . .!"

Here are Some Purchasing Tips

I purchased my unit from a reputable dealer who would back his guarantee and provide good service when I needed it. Advisedly, you should look around before you decide which set you want to purchase, for there are dozens from which to choose. Use the following list as a guideline:

(1) What do I want from my CB? What uses will it serve?

(2) How much money do I want to spend?
(3) Talk to a number of your friends who have CB's.
(4) Should I buy a mobile or base?
(5) Don't get in a hurry!

Whether you plan to use your CB for travel, business, or just for a hobby, you will have a good investment. But you need to decide what your plans are and what you hope to gain from your purchase. Later on in this book I will discuss the many and varied uses of your CB.

Money is a factor when you purchase anything, but especially a CB. Most units are about the same after you open them up—provided they are "type accepted" by the FCC. Transmitter type acceptance means that the manufacturer has submitted documentary evidence to the FCC which certifies that the transmitter meets the technical standards established by the FCC. Look for the identification plate, easily recognized, either on the back or bottom of the radio.

The "frills" run the cost up. Such accessories as external auxiliary speakers (P.A.), built-in clocks, and various meters can double and sometimes even triple your initial cost. You need to decide what you can afford to spend (a range) and stick with it, for you can purchase the "squawk box" anywhere from about $70 to $750, depending on the brand. *Remember*—the antenna is usually not included in the cost of the radio, and the antenna is so important, for it helps determine just how effective your set will be.

Before I finally decided to buy my CB, I did quite a bit of shopping around in the electronic stores and elsewhere. You now can purchase equipment and accessories at department stores, discount outlets, furniture stores, and even from some TV repairmen. I talked with friends who had units and asked about their likes, dislikes, and where they bought their sets. Most were of the opinion that you need to buy from an outlet that will provide good service and which has a licensed FCC operator employed to answer questions and give advice on repairs and accessories. All of the stores are in the business for profit, but not all are concerned with the satisfaction of the customer.

Your first inclination in purchasing will probably be to get a mobile rig for your car or truck. However, don't rule out the possibility of a CB that can be changed over to become a base station. Almost all can, but some are more versatile than others. Some sets are equipped with dual-voltage supply that will allow them to accept both DC and AC (household) current without having to add a separate piece of equipment. If your CB is not equipped in this matter, you will need to get a power supply which converts 120-volt AC to 12-volt DC.

The point I am trying to make is that you can use your rig in both places, enjoying the benefits of both mobile and base communications.

The last tip is *don't get in a hurry* and end up

buying something you may be unhappy with in a few months. Taking a little extra time to find out the best buy for your money and your needs will save some headaches down the road. Don't be surprised if the brand and type you choose is out of stock. The more popular models usually have to be ordered, but the wait is worth it. Another important thing to remember is to ask for a demonstration. Any reputable dealer will gladly take time to explain about any particular set and demonstrate it for you.

A typical "base station"

17. The CB Unit

Your CB is called a "transceiver," for it receives as well as transmits sound. When connected to a power supply and equipped with the right kind of antenna, the transceiver will send out and receive transmissions from distances of approximately ten-fifteen miles, depending on the terrain and atmospheric conditions.

The unit is made up of interconnected transmitter and receiver circuits. The transmitter circuit converts the spoken message into electrical energy and sends it out through the antenna as a radio signal; the receiver converts the radio signal into sound waves which are heard as voice. The FCC allows a 5-watt input and 4-watt output. Compare this to Ham operators (up to 1,000 watts) and radio stations (50,000 watts or more) and you begin to understand the limited range of the CB. Class D limits the CB range to 150 miles, but ordinarily it is impossible to transmit that far unless a "skip" occurs, or special equipment is connected to the CB unit.

A "skip" is that situation which takes place when certain atmospheric conditions make it possible for signals to be picked up hundreds and even thousands of miles away. Skip is evidenced more frequently during the daytime than at night because the presence of the sun during the day causes layers of particles in the ionosphere to become electrically charged. These charged particles act like mirrors and reflect back to earth. If skip transmission interferes with your modulation, it may be necessary for you to wait until atmospheric conditions become more conducive to straight transmission.

Recently, while pulling into my driveway, I heard a conversation between two "good buddies" in New Mexico. That is a pretty good jump from Louisiana! I didn't break in on the modulation. If I had, I would have been breaking the law because the FCC (Federal Communications Commission) prohibits talking "skip." I did, however, enjoy listening.

Linear amplifiers and large antennas are examples of special equipment which boosts your signal. However, it should be pointed out that linears, like "skip" talking, are illegal because they interfere with local communications.

18. Your Antenna Is Important

I repeat: *remember the importance of your antenna.* The antenna determines in large measure the effectiveness of your modulation. You may spend $200 or $300 on a radio and experience poor transmissions if your antenna is inadequate. Mobile units and base units require different antennas. Those of larger more elaborate design are needed for base operations. The mobile unit may operate satisfactorily with one of the following:

(1) *Types of Mobile Antennas*

A. *Base Loaded*–This antenna is correct in electrical length, but shorter than the whip. Do not get one less than 42 inches in length. The strength of the "base loaded" is in the concentrated coil. This type aerial will work best in the center of the roof, or the middle of your trunk.

B. *Whip*–The "whip," a nine-foot aerial, provides optimum electrical body for resonance, but presents problems in mounting. Those mounted on roofs frequently bump tree limbs, garage ceilings, and other obstructions. Those attached to bumpers sometimes experience radiation blockage because of ground plane interference.

C. *Dual Antennas*–No doubt you have seen the 18-wheelers with "dual ears." Perhaps you have even noticed them on "country Cadillacs," motor homes, and buses. The two antennas work together to produce an excellent signal. The distance between the antennas is important in determining the directions the signals travel. Antennas positioned less than eight feet apart travel best sideways. Those separated by eight feet or more move out better anteriorly and posteriorly.

D. *Fiberglass*–This variety is the newest on the market and offers strength and durability. They are usually pre-adjusted for resonance and are available in both the whip and four-foot length.

E. *Metal*–These come in numerous lengths and styles and work well with the spring base roof mount. Most CB manufacturers recommend that you not use aluminum material.

(2) *Base Station Antennas*

When considering a base station antenna, it would be advisable to talk to a local dealer about the type best suited to your part of the country and your area of town.

There are basically two types of base antennas: directional (beam) and omnidirectional. The omnidirectional transmits equally in all directions. The beam aerial focuses energy in only one direction at a time, but achieves greater distance than the omnidirectional. FCC rules and guidelines should be followed in setting up both of these antennas.

The amount of gain the antenna has will be an important consideration.

> "The simplest antenna, a regular ¼ wave vertical, has no gain. Other antennas are measured in how many times better than this they are. Let's say that you just bought a new antenna that has a gain of 3 dbs (db stands for decibels, which is what gain is measured in). When you hook up your four-watt rig to it, you'll increase your effective power past what your ¼ wave vertical antenna could put out by itself. A 3 db gain is approximately a doubling of power."[2]

Other distinctives to examine before buying are durability of construction, weight, and details of assembly.

(3) *Talk With Others*

After I got my unit and was operating, I oftentimes modulated with those who had a different type antenna. I knew that I was going to be getting another one for my second car and wanted to get the best. The placement on the car is most important and I wanted to make the most of the experience of others.

(4) *Where's the Best Place for the Mount?*

The best way to answer that is with the following illustrations. Mobile antennas are omnidirectional, but because the car or truck contributes to the overall ground plane, they become directionally stronger depending on where they are placed on the vehicle.

Antennas mounted on the center of the *roof* and the *rear bumper* produce signals in this manner:

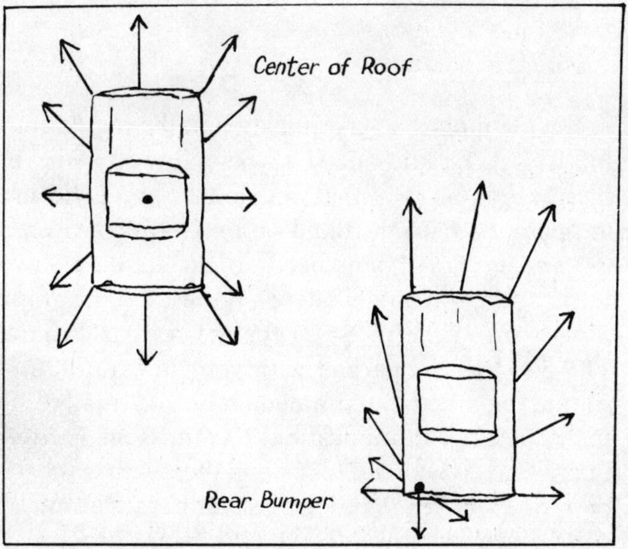

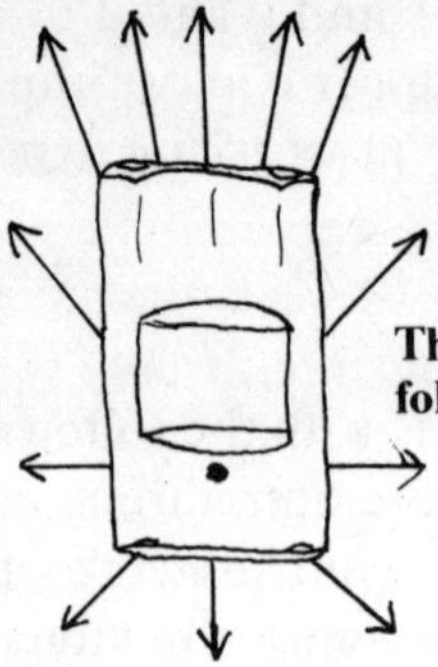

The popular trunk mount will give the following effect.

19. Get Your License

Let me mention very quickly that the regulations for Class D radios need to be followed. One of these regulations is that you must have a license to operate a Citizens Band radio. Getting a license is very easy. All you need to do is secure two forms–505 and 555B. There's no test to take–only a very small fee of $4.00 for a five-year license. The 505 form is packed with your new radio and should be filled out immediately and mailed to the Federal Communications Commission, Gettysburg, Pennsylvania 71326, together with a check or money order. Your license and call numbers will be sent to you as soon as they are processed.

With applications running 300,000-plus each month, the waiting period is about eight-ten weeks.

It has been estimated that over 30 percent of the approximately fifteen million operators are not licensed. Suffice it to say that if you plan to get a CB and use it as a Christian should use it–get a license! Be in accordance with the rules and abide by the spirit of the law. When you sign your license application, you are pledging that you will do this. Also notice in the bottom left hand corner of Form 505 the item relating to Part 95 of the Commission's rules governing the Citizens Radio Service. Order this section and be familiar with the do's and don't's, but be prepared for a lot of technical data.

Rules and regulations may seem like a pain in the neck, but they are for the benefit of all CB'ers. On several occasions I have heard operators caution other operators on a technicality of the rules. The five-minute conversation time limit is often abused. Cooperation is the key in sharing the channels–always try to be courteous and considerate.

20. What About Installation?

Installing your CB can be simple or complicated, depending on where you want to place it. In most instances, it is merely a matter of following directions and the advice of persons who have had experience in installation.

For an extra $20-$25 you can get a professional job from your dealer. I paid to have my first one installed because I wanted a roof mount antenna, which required drilling a hole and fishing the antenna cable from behind the ceiling upholstery. This process involved equipment I did not have. Professional assistance was more convenient in this case. I installed the other one myself with the help of a friend and had no problem at all. Here is a brief checklist if you plan to make the installation yourself:

(1) *Decide Where You Want the Unit.* This is usually determined by the individual. I personally like the controls to be within easy reach of my left hand. Most installations are toward the middle of the car, depending on gearshift knobs, air conditioning, etc. In compact cars, it is often necessary to place the rig to the left of the steering column. (See pictures) Driving safety is also a factor to

consider since it is best not to lose sight of the road while reaching for your mike. That may lead to a permanent 10-7!

(2) *Check the Ground System.* Most cars have a negative ground which means that the negative battery terminal is grounded to the frame. However, this may not be the case, particularly with foreign conveyances. Urgent! It is wise to be certain before you begin the installation process.

(3) *Negative Ground.* The best and easiest installation here is to run your power wire to your fuse box and insert the frayed end behind one of the fuses. Install an in-line fuse to prevent damage to your unit in the event of a short in the car's electrical system. If you are anchored to metal, then no further grounding is needed. On the other hand, if you are anchored to plastic, you will need to ground the set to metal somewhere beneath the dash. (Note: Choose the radio or ignition fuse so that your set will turn on and off with your key).

(4) *Positive Ground.* This type of installation is much more involved, and the best advice here is to consult with your dealer. Chances are, unless you have a semi or truck, your car will be negative ground.

(5) *6- or 12-Volt.* CB's are designed to be used for 12 Volts DC or 120 Volts AC. If your car is 6-Volt, you will need to buy a converter.

(6) *Anti-theft Devices.* Many types are on the mar-

ket, and they do everything from activating your horn to producing a shock. One operator was ripped off so many times, he decided to put his five pet snakes in his car when it was unattended. Signs were posted on the windows announcing a warning, "By the way the 'snakeman' no longer loses his radios."

21. Now, You're Ready to Modulate

One of the first things you'll want to do is to get yourself a "handle." A friend recently came into my office with a list of 214 "handles" that she had collected on a trip to New Mexico. Almost every conceivable name was on the list from "Adam's Housecat" to the "Blue Zebra."

In selecting yours, choose something that is understandable and easy to pronounce. Be careful to choose one that will not cause embarrassment. For example, I recently communicated with the "Bullshipper"!! Needless to say, I enunciated cautiously during my modulation with him. Everyone uses a "handle" so decide on one before you begin. You will enjoy your conversations much more and feel at ease doing so.

Looking for a handle that no one else has may be a tall order to fill, but it can be done. "Rubber Ducks" and "White Knights" are found by the dozens due to the popularity of the hit records. Try to be creative, and you'll come a lot closer to getting a handle that is clever and new to the scene. By the way, don't feel embarrassed when you have to change–it happens quite often! My handle is "Louisiana Canine," and I have yet to hear anyone else with this same handle. By adding the state to it also cuts down on duplication. There may be a "Texas Canine" or a "Carolina Canine," so the state boundary makes it unique. Many operators use a handle that may relate to their occupation. Some of these are "Pipeman" (Plumber), "Keymaker" (Locksmith), or "Inksetter" (Printer). You will have fun deciding which "handle" suits YOU best!

In a later chapter, I will deal with handles and the possibility of using a Christian-related handle.

Controls and Buttons

Most rigs have the same basic controls and knobs. Remember that it doesn't take a Ph.D. to operate a CB, but you need to know a few things. The basic controls are an on/off/volume switch, channel selector, squelch control, and mike. If you can find these, and your radio is installed properly, you're in business.

The squelch control is probably the only one of the four that may require some explanation. Basically, squelch takes care of background noise and static. Turn the control all the way over to the left, then begin turning it to the right until you get just past the noise and static level. Now, you can monitor whatever channel you select without the constant static. But do remember this fact: the further you move to the right of the noise level, the less range you have. Weak signals require that you move back into the noise level.

Don't Be Afraid!

If you like, just listen a while before you attempt to modulate with a "good buddy." The jargon is different, and it sometimes takes a while to get your ear accustomed to this type of conversation. Try to pick up handles, and see if you recognize a few of the 10-signals that are frequently used. It may take a while, but your fun is just beginning.

Which Channel?

There are twenty-three channels from which to choose unless you purchased one that contained less. I don't advise doing this unless your 6-channel or 3-channel unit is an additional rig for a second car or truck. Communications with other family members is then simplified by predetermining which channel you will use. Just remember that the channels are crowded, and you will be handicapped if you have less than twenty-three. Then, too, FCC has recently released a report that manufacturers will be authorized to offer seventeen additional channels on new CB's beginning January 1, 1977, bringing the total to forty. But don't worry about your twenty-three-channel unit getting out of date–it won't. The additional channels will simply make it easier to move out of the presently congested twenty-three. Contact channels like 11 and 19 will remain popular for years to come.

Most truckers use the same channel all over the country–19. Traveling between cities you will more than likely want to get on this channel to stay abreast of the latest road conditions. Channel 11 is a contact channel, and it is suggested you move to an open channel after contacting your party. In areas that are heavily populated with "ears," the "calling channel" is not used as much. When this overcrowding occurs, locating any open channel is like finding a pot of gold. Channel 9 is for emergency use only! Use this in communications involving immediate safety of life or protection of property and/or assistance to motorists. But it should be used only for that purpose. The other twenty-one channels are open for anyone to use.

The Channel Break!

After deciding the channel, and if it is clear, take the big step and "break, break!" Here's an illustration.

> "Break, Break, 19, this is KTP1206, the 'Louisiana Canine,' looking for a radio check."
>
> "Go ahead, 'Louisiana Canine,' you got a radio check."
>
> "What's the handle on that end?"
>
> "You got the 'Leaping Lizard' on this end."
>
> "10-4, Mr. Leaping Lizard, what am I putting on you?"
>
> "About a 9-plus, Mr. Canine,–coming in wall-to-wall!"
>
> "Yeah, that's a big 10-4, 'Leaping Lizard,' appreciate the report. I'll just move on off this 19 and let somebody have it."
>
> "10-4, Canine, enjoyed modulating with you,

have a good day and a better one tomorrow. This is KHC71360, the one 'Leaping Lizard' clear."

"KTP1206, 'The Louisiana Canine,' I'll be 10-10 and listening in."

You will learn a lot by listening to others, but there is no need in being skittish or nervous about breaking the channel. Another illustration shows the correct procedure for contacting the emergency Channel 9, in most areas manned by REACT teams twenty-four hours a day.

"Break Channel 9, this is KTP1206 with emergency information."

"KHC7011, Cenla REACT, proceed with emergency information."

"Got a 10-33 with injury two miles north of Hoffman Bridge on US 75. Request police and ambulance."

"10-4, KTP1206, information received and will be forwarded to proper authorities. Any other information."

"Negatory, KTP1206, clear."

"KHC7011, Cenla REACT, clear."

Remember that channel 9 is all business and should never be used for mundane conversation. Recently, while monitoring channel 9, a breaker came on wanting a 10-36. Almost immediately, the REACT monitor came back with a 10-30 (does not conform to FCC rules). He asked courteously that the breaker go to one of the other twenty-two channels for non-emergency information.

How Far Can I Talk?

Many factors influence how far your unit will transmit (Atmospheric conditions, tall buildings, quality of receiver and antenna, weather, etc.). I have modulated over distances of five-ten miles when I was to the right of the noise level (squelch) and up to fifteen miles when to the left of the noise level. However, I can recall one incident when I reached out much further.

I was traveling east on US 190 and began modulating with "Kojak," a local police officer in a small town just east of Opelousas. It was late at night, and very few folks were on the channel. I asked if he would mind giving me a radio check at one-mile intervals. Kojak was glad to accommodate since he wanted to check his unit out, too. The signal was faint, yet discernable, at twenty-two miles! Evidently, the conditions were just right, for it is highly unusual to transmit much over fifteen miles (mobile to mobile), unless you catch a skip.

Now, base stations are different. Those with directional beam antennas will reach out fifteen—twenty-five miles fairly consistently but again depending on weather, obstructions, and quality of receiver and antenna.

22. Trouble Shooting (It Quit Working)

"The only thing Jim knows about his car is where to put the key–if that doesn't work, he forgets it." A lot of CB operators fall into this category when it comes to their radio. If it doesn't come on, or if they can't at least get some static, they are out of business.

Anytime you have trouble, the best rule to follow is: "get professional help." Unless you have a First- or Second-class Radiotelephone Operator license, you are not legally qualified to repair the power circuits of a CB radio. But before you go to your dealer or an experienced serviceman, check out some possible trouble spots.

(1) Check your *connections.* If a wire is frayed or is not making good contact, your set will be affected.

(2) Check your *fuses.* The in-line fuse in the power wire is the most frequent one to blow. Replace it with a 3-amp fuse–don't use tin foil!

(3) Check your *ground* for proper contact.

(4) Check your *mike connection.* Continually stretching the microphone cord causes pressure on the connection.

23. Various Uses of Your CB

In a previous chapter, I discussed primarily the Christian uses of CB, giving various testimonies and personal experiences. Now, let us consider the other uses of your radio. A complete listing is almost impossible, because new uses are discovered each day either on purpose or by accident, but this list will help.

(1) *In Business*—More and more businessmen are finding that CB radio can help save time and gasoline, especially when the job calls for service or delivery. Business associates stay in touch.

(2) *Family Use*—Get messages to your wife (husband), or stay in touch with friends and neighbors. You name it!

(3) *Salesmen*—Efficiency and economics are two factors considered by persons in sales. The satisfied customer means another sale in the future. CB can help do the job more effectively.

(4) *Public Service Use*—Most city and county service agencies use two-way radios, but a greater number are moving to CB due to the lower cost. This is especially true in the smaller towns. Also, volunteer fire departments are finding that CB's

can get the job done when getting out the word to volunteers.

(5) *Medical Personnel*—Doctors stay in touch with their officers. Although I don't have an incident to support this, I'm sure there have been times when doctors or other medical personnel have responded to a call for help over CB (Auto wreck, pedestrian injured, etc.).

(6) *On the Farm*—No matter what size "spread" you have, there just aren't many telephones conveniently placed unless they're in your car or truck. And that's usually a CB! Many farmers have begun to see the value of CB in tractors in order to be in touch with workers.

(7) *For Area Security*—Many large subdivisions or exclusive housing areas have begun to hire security police to try and cut down on thefts and vandalism. Walkie-talkies are widely used for this purpose.

(8) *Inter-City Travel*—Find out about road conditions, hazards, and accidents as you travel. And a CB is the best road map you'll ever have. Truckers usually know all the short cuts. When you roll into a town and need to get directions, help is available almost immediately. You can even find out the best restaurants.

(9) *Family Vacations*—I knew a couple that had to take two cars on vacation because they had nine children. They stayed in touch with their CB radios.

Your trip will be safer and more enjoyable with a "rig."

(10) *Hiking*—explained earlier

(11) *Safety*—More and more women who travel alone are getting CB's as a safety device. Not long ago I talked with a lady who got a CB as a result of an incident which could have ended her life. She was traveling on a little-used state highway when a car repeatedly tried to force her off the road. She was able to get away, but a CB sure would have been helpful in the situation. Then she could have been able to call for assistance. My wife, Jo, on occasion travels alone and just knowing the CB is in the car is a comfort to her and to me! Nurses, shift workers, and other ladies who have to drive at night ought to consider getting a CB.

(12) *Emergencies*—explained earlier

(13) *Fishing and Hunting*—I don't have a CB hookup for my pickup, but I've wished for one several times when stranded due to getting stuck or lost! Fishermen are extensively using CB's to find out where the "hot spots" are. Deer hunters keep up with where the dogs are running.

(14) *Forestry Service and Ski Patrols*—CB is helpful to search parties when campers or skiers get lost. Ski patrols can get doctors' orders in the slope before moving an accident victim.

(15) *Surveyors*—self-explanatory

(16) *Construction Crews*—self-explanatory

I'm sure, after you have used your CB for a while, you'll discover many more "uses" for this unique instrument. The sky is the limit!

The classic "18-Wheeler" With "Dual Ears"

24. Summary

In conclusion, there are certain essential details which must be known, whether you are making CB a hobby or using it for your livelihood or a combination of both. The following questions and answers are prepared by the FCC to provide these essential details as a "quick reference."

Q. What is the Citizens Radio Service?
A. The Citizens Radio Service (CB) is a short distance, two-way radio communication service intended for the personal and business use of licensees. The official name for CB is *Class D Citizens Radio Service.*

Q. Do I need a license in order to use CB radio?
A. Yes, operation of a CB transmitter without a valid license will subject the user to a $10,000 fine and/or one-year imprisonment.

Q. Who is eligible for a CB radio station license?
A. Any person 18 years of age or over who has a mailing address in the United States, except a person who is a representative of a foreign government.

Q. Can I obtain a temporary permit to operate my CB station while waiting for issuance of my regular

license?
A. Yes. You must first obtain a copy of FCC Form 555-B entitled *Temporary Permit Class D Citizens Radio Station.*

Q. What is meant by a call sign?
A. Every CB radio station is issued a call sign by the FCC. These call signs currently consist of three letters and four numerals, for example, KYP-2965.

Q. How must I identify my radio transmissions?
A. Identify your radio transmissions with your own FCC issued call sign before and after each transmission. "Nicknames" or "handles" may also be used to identify your radio transmissions *provided they are accompanied by the FCC assigned call sign.* It is not necessary to transmit the call sign of the station with whom you are talking.

Q. What are the important factors to consider when using the CB channels?
A. Cooperate with all CBer's to the fullest extent possible in sharing the CB channels–always try to be courteous and considerate when using a channel. In order to assure that all CB operators will have an equal opportunity to use the frequencies, radio communications between CB stations (interstation) must be limited to no longer than five continuous minutes to be followed by a silent period of at least one minute.

Q. Who is permitted to operate a CB transceiver?
A. If the CB station is licensed in the name of

an individual, then the licensee, members of his immediate family residing in his household, and his employees while acting within the scope of their employment may operate the station.

Q. What is the average "communication range"?
A. Since there are many factors that influence the "communication range" (such as terrain, atmospheric conditions, weather, physical obstructions, antenna height, quality of transceiver and antenna, etc.), it is impossible to state with any degree of certainty exactly what the range will be. The average "communication ranges" are as follows:

Mobile to Mobile–5 to 15 miles
Base to Mobile–10 to 25 miles
Base to Base–15 to 30 miles

Q. May a CB'er use his CB radio to transmit over long distances? (i.e. distances greater than 150 miles)
A. No. CB'ers may not use CB to transmit over 150 miles. Sometimes CB transceivers pick up stations hundreds and even thousands of miles away by means of "skip." Skip occurs when CB radio waves hit the ionosphere and bounce back to earth at least several hundred miles away from their point of origin. Because talking "skip" increases local congestion as well as in the "skip" area, talking via "skip" is prohibited.

Q. What is the best way to select a CB transceiver?
A. Before you purchase a CB transceiver, make certain it is type accepted by the FCC. Any FCC

office or CB equipment dealer may help you determine whether a particular CB transmitter is type accepted.

Q. May I repair my own CB radio transceiver?
A. Yes, but only if you possess a First- or Second-class Radiotelephone Operator license or if you do the work under the supervision of the holder of such a license. It is recommended that repairs be made only by a professional technician who has the necessary license and test equipment.

Q. Are power amplifiers permitted to be used by CB'ers?
A. No. The use of power amplifiers, power boosters, or linear amplifiers is strictly prohibited. Use of these illegal devices can cause serious radio frequency interference problems to TV sets, hi-fi equipment, electronic home entertainment devices, public address (loud speaker) systems, other CB radio stations, and stations in other radio services. The maximum output power for a conventional, AM, CB transmitter is four watts.

Q. What are the two CB antenna types and what are the maximum permissible heights?
A. Two types of antennas used in the Citizens Radio Service are Directional and Omnidirectional antennas.

Q. How may I keep informed about the CB Radio Service?
A. You may obtain the FCC Rules and Regu-

lations (Volume VI, Part 95) governing the CB Radio Service from the Superintendent of Documents, Government Printing Office, Washington, D.C. 20402, for a cost of $1.50. Copies of these rules are also available at many radio equipment stores at a nominal cost. There are also available from many radio equipment suppliers a variety of excellent publications concerning the CB Radio Service.

Q. What can CB'ers do to assure maximum returns from their CB radio station investments?

A. CB'ers should do the following:

(1) voluntarily operate their stations in full compliance with the FCC Rules, and

(2) encourage other CB users to do likewise.

PART FOUR

Handles, Jargon, and Signals

Dr. Landrum P. Leavell, president of New Orleans Baptist Theological Seminary, became convinced of the need for a CB radio when he had car trouble on a road between Sorrento and LaPlace (Louisiana). After obtaining a CB, he decided to use the handle, "Preacherman."

A young minister made the comment to Dr. Leavell that no one would ever modulate with him with "that handle." The minister's fears were unfounded as illustrated by this incident Dr. Leavell shares.

> "My handle is 'Preacherman.' I had been in a convoy coming down Interstate 55 from Memphis to New Orleans one Sunday afternoon, May 30. One of the fellows in the crowd had the handle "Mountainman." He was turning west on I-12 at Hammond, and before he got off I-55, he asked me if my handle had any relationship to my occupation. I gave him a Big 10-4! He immediately came back and asked me if I would pray for him and his family, that they might have a safe journey on their vacation. I assured him that I would. I know for a fact that there were at least eight or ten others who heard the conversation."

Handles can be helpful to the Christian wanting to share his faith in Christ, as evidenced by the above testimony. When I hear a person's handle, I immediately try to figure out his occupation or pastime that might be related to the handle. I do this evaluation with all handles, but my "ears" especially perk up when I hear ones with "Christian distinctives."

Federal law requires that all CB'ers use call letters and become licensed, but most use call letters along with handles. A person with a Christian-related handle can usually expect one of two things to happen after transmitting:

(1) There will be an immediate conversation flow which may lead to the possibility of sharing a testimony or spot announcement.

(2) The "cold mike" reception, which might be termed a negative conversation flow (I don't think I want to modulate with you).

Bill Emmitt, associate pastor at Istrouma Baptist Church in Baton Rouge, Louisiana, told me he had a chance to modulate with a couple of Christians on La. 71. Their handles were "Child of the King" and "Soldier of the Cross." Bill said, "Our conversation must have been a witness to all who had their ears on, for we had a great time praising the Lord."

Here is a list of handles with "Christian distinctives." You might find them useful as you choose your handle or modulate with persons who have similar identifications.

Child of the King
Glory Road
Good News
Soldier of the Cross
Glory Trip
King's Kid
Glory Train
Goliath
Preacherman
Padre
Jubilee
Dry Bones
Joe Baptist
Lady Baptist
Jethro
Widow's Mite
Footstool
Doubting Thomas
Brotherly Love
Golden Rule
Good Tidings
Good Influence
Georgia New Day
Heaven's Peace
Sword Bearer
Deacon
"Deak"
Faith Seeker

Lake Missionary
Music Man
Colorado Tune Heister
Rollin' Chaplain
Disciple
Sharing Man
The Cross Man
The Rev.
Good Samaritan
Noah's Ark
Servant Boy
John the Baptist
The Dove
Dr. Luke
Priceless Pearl
Loaves and Fishes
Rich Ruler
Good Shepherd or Sheepherder
True Joy
One Way
The Prophet
Fisherman
Burden Bearer
Mercy Train
Cross Bearer
Lost Sheep
Truth Seeker
The Bishop

Chaplain
Blue Monk
Good Steward
Cajun Rabbi

Guy Ward of Salt Lake City, Utah, relates this interesting bit of information on his handle:

"My CB handle is 'Joe Baptist' and my wife is 'Lady Baptist.' In this part of the country just the term 'Baptist' is enough to initiate curiosity. My wife and I chose these handles as testimonies in themselves. For instance, my wife inadvertently made contact with a lady in Magva, Utah. The lady had 'Baptist' in her background. A witnessing and sharing opportunity resulted."

A central Louisiana pastor was monitoring his CB late one night and related the incident that involved an 18-wheeler and a four-wheeler. A lady breaker was in the four-wheeler advertising "her services" to anyone who would listen. The 18-wheeler answered her with the statement. "I don't think my Father would approve of such activities." "Temptation" (her handle) replied, "Your father wouldn't mind." "Yes, he would," said "Child of the King." "My Father cares about me, and he cares about you. He's a King."

"Who *is* your father, anyway?" asked Temptation.

"He is the King of kings and Lord of lords!"

The pastor said that the channel went silent for a full thirty seconds. No one said anything! Child of the King then proceeded to share his testimony of how he knew that he was a "child of the King."

JARGON

There was a time when I was certain I had a photographic memory. I could memorize page after page of printed material and etch into my grey matter unbelievable details about people's names, clothing, furniture arrangements, and an assortment of other things. If a photographic memory, indeed, was the case, then I am equally as certain at this point in my life that the film has been overexposed. Not only do I fail to recognize the names and faces of most people outside my immediate family, I have to look up my own phone number when I want to call home!

Perhaps some of you readers can relate to these infirmities. If so, then this section is for you. The drawings you see are visual images I have created to help me remember the meaning of some of the CB lingo. Far be it for me to memorize all that jargon, but I have included a fairly complete list of signals and slang for those of you who might aspire to commit it all to memory. They're not listed in strict alphabetical order.

CB JARGON

Addict . . . CB operator who stays on the channel almost constantly.

Affirmative . . . Yes.

Afterburner . . . An illegal amplifier. Power booster.

Alligator . . . All mouth and no ears. Never listens. Only talks.

Anchor . . . Last truck or car in a caravan or convoy.

"Bear in the Air"–police in helicopter or airplane

Anchor Man . . . Operator of a base unit (station).
Antenna Farm . . . Commercial TV or radio broadcaster.
Apple Barrel . . . Meeting of CB addicts.
Astrodome City . . . Houston, Texas.
Audio . . . Voice.
Adios . . . Going off the channel. Going 10-7.
Back Door . . . Usually the last CB'er in a convoy.
Back Down . . . Slow down. Stop modulating.
Barn . . . Garage for truck.
Base Station . . . CB unit at a fixed location.
Baby Buggy . . . School bus.
Bear . . . Policeman. Highway Patrol.
Back . . . Asking for a reply as in "come back."
Bear in the Air . . . Police helicopter or airplane.
Back Off the Hammer . . . Get your foot off the gas.
Bear in the Bushes . . . patrolman hiding.
Background . . . As in background noise (static, interference).
Bag Hound . . . Baltimore, Maryland.
Banana . . . A CB'er who likes to hear himself talk.
Beaver . . . Girl CB modulator.
Big 10-4 . . . Whole-hearted agreement or affirmative.
Break . . . Channel request. (Break, Break, One-Nine)
Brush Your Teeth and Comb Your Hair . . . Go to legal speed limit.
Back It On Down . . . Slow down.

"Boomer"–any vehicle carrying explosives

Backstroke . . . Return trip.
Bad Scene . . . Crowded Channel.
Bar City . . . Forest City, Arkansas.
Big A . . . Amarillo, Texas.
Big Apple . . . New York City.
Big D . . . Dallas, Texas.
Big Ears . . . Getting good, clear reception.
Big M . . . Memphis, Tennessee.
Bite in the Britches . . . A speeding ticket.
Bleed Over . . . Overpowering transmission which is heard on other channels.
Blowing Smoke . . . A strong, clear signal.
Blood Bucket . . . Ambulance on the move.
Bodacious . . . usually refers to a very strong signal.
Boomer . . . Any vehicle carrying explosives.

Box on Wheels . . . A hearse.

Breaking Up . . . signal is fading in and out because of interference or distance.

Bubble Gum Machine . . . Highway patrol or policeman with lights mounted on the roof.

Call Letters . . . identification letters issued by the FCC.

Capital J . . . Jackson, Mississippi.

Channel Hog . . . CB'er who constantly goes over his five-minute allotment for conversation.

Chicken Coop . . . Interstate Commerce Commission (ICC) weigh station.

Choo Choo Town . . . Chattanooga, Tennessee.

Christmas Card . . . Getting your CB license.

Cinderella City . . . Disneyland, California.

Cliff Hanger . . . Two-lane road with lots of curves and inclines. Usually relates to a mountain road.

Coax . . . Coaxial wire that runs from antenna to transmitter.

Coke Stop . . . Stop for Mother Nature.

Cup of Mud . . . Cup of Coffee.

Convoy . . . A number of vehicles traveling in the same direction using CB's to stay in contact.

Cooked Goose . . . A person who has been caught speeding by a highway patrolman.

Country Cadillac . . . Pickup truck.

Cow Town . . . Fort Worth, Texas.

Clean Shot . . . No radar units. Road is in good condition.

Clear . . . Term used in signing off the air.

County Mounty . . . County sheriff or patrol car.

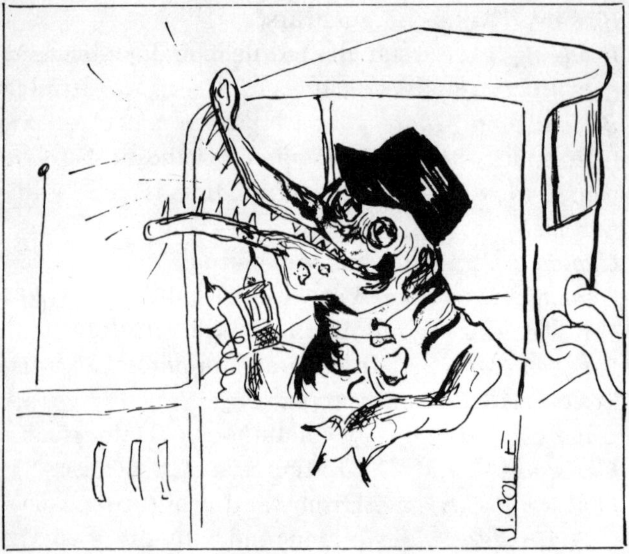

"Alligator Mouth"–person who hogs the channel

Dead Pebble . . . Slow vehicle.

Dent Collector . . . Reckless driver. Heavily dented vehicle.

Diesel City . . . Detroit, Michigan.

Dig You Out . . . Understanding a transmission even though there is a lot of interference on the channel.

Do You Copy? . . . Can you understand me?

Don't Feed the Bear . . . Be careful not to get a speeding ticket.

Drop the Hammer . . . Give it some gas. Accelerate.

Double L . . . Land Line. Telephone call.

Double Nickel . . . 55-miles-per-hour speed limit.

Ears . . . Refers to antennas.

Eye in the Sky . . . Airplane checking for speeders.

Eighteen- (18-) Wheeler . . . Big tractor trailer truck.

Eights (3's and 8's to you) . . . Greetings. Best of luck to you.

Eyecatcher . . . Good-looking CB'er.

Feather Foot . . . A person who continually speeds but never gets caught.

Fat Load . . . Large, extra-heavy load.

Feed the Bears . . . Pay a speeding ticket.

Final Edition . . . Going off the air at the end of the day or trip.

Five by Five . . . Good, strong signal.

Five Watts . . . Maximum legal input for a Class D CB transceiver.

Flag Town . . . Flagstaff, Arizona.

Flapper . . . CB antenna on a mobile unit.

Flip Side . . . The return trip as in "flip flop."

Foot in Mouth . . . A modulator who always seems to get into trouble because he talks too much.

Fluff Stuff . . . Snow.

For Sure . . . Agreement. 10-4.

Four-lane Parking Lot . . . Overly crowded four lane road.

Four-Ten . . . Another way of saying 10-4.

Four Wheeler . . . A vehicle with four wheels. Usually a car.

Front Door . . . The lead vehicle in a convoy.

Funny Book . . . Trip log book for truckers.

Gateway City . . . St. Louis, Missouri.

Gentle Ben . . . Any police officer or patrolman.

Getting Out . . . Strength of transmission. (How am I getting out?)

Give a Shout . . . (Tell the "Rubber Dog" to "give me a shout.")

Go Ahead . . . You can talk; the channel is clear.

Go, Breaker . . . Same as above.

Good Buddy . . . Term used to describe another CB'er.

Good Numbers . . . Best of luck to you. Best wishes.

Got A Copy? . . . Did you hear my transmission?

Grandma . . . Vehicle is in neutral.

Green Light . . . Everything is clear.

Green Stamps . . . Money.

Ground Clouds . . . Foggy conditions exist.

Gutter Mouth . . . An operator who uses profanity and obscenities on the air.

Haircut Palace . . . Low clearance. Underpass.

Ham . . . Licensed amateur radio operator.

Hammer Down . . . Accelerate as in "let the hammer down."

Handle . . . Nickname.

Handle Grabber . . . A person who adopts a handle someone else has.

Handle Hog . . . Operator who changes "handles" frequently.

Hat Box . . . Gear Box.

Hip-pocket Wonder . . . Hand held CB or Walkie Talkie.

Hog City . . . Little Rock, Arkansas.

Hole in the Wall . . . Underpass or tunnel.

Home Plate . . . Where you live.

Home-20 . . . Same as above.

Hot Lanta . . . Atlanta, Georgia.

Hydroplane . . . Wet, slick surface causes fast-moving vehicle to leave the road.

Ice Box . . . Truck or trailer that carries a refrigeration unit.

Ink . . . Oil or grease.

Invitations . . . Police traffic citations or tickets.

Jack Rabbit . . . Fast-moving vehicle.

Jamboree . . . Gathering of CB'ers in regional meetings.

John Law . . . Any police officer or patrolman.

Key . . . Lever or button on microphone.

Keep the Shiny Side Up and the Dirty Side Down . . . Make your trip a safe one.

Keep Your Wheels Spinning and the Beavers Grinning . . . Sign off. Closing out of a conversation.

Kicker . . . Illegal amplifier.

Knuckle Buster . . . Street fight. Brawl.

K-Town . . . Kansas City, Missouri.

Land Line . . . Telephone.

Lettuce . . . Green Stamp. Money.

Listening In . . . Monitoring.

Load of VW Radiators . . . Empty truck.

Local Yokel . . . Local patrolman.

Long Distance Call . . . Break for a specific CB'er. "I need to contact 'Green Snake.' Anybody heard him?"

Lookin' Good . . . You are coming in strong.

Lucy Anna . . . Louisiana.

"Squawk Box"–CB unit

Lumber Bumper . . . Log Truck.
Magic Eye . . . Radar Machine.
Make the Trip . . . After moving to another channel, resume contact by asking: "Did you make the trip?"
Mama . . . Wife. XYL.
Mama Bear . . . Patrolwoman.
Man in Red . . . Fireman.
Mardi Gras Town . . . New Orleans, Louisiana.
Meat Wagon . . . Ambulance.
Mike Fright . . . Operator who is afraid to talk or won't speak loud enough into his mike.

Mike Marker . . . Milepost on highway.

Mobile . . . CB unit operated from a moving vehicle.

Mobile Parking Lot . . . Truck carrying autos.

Modulate . . . Talking or conversing over a CB.

Motion Lotion . . . Diesel or gasoline.

Motor Boater . . . Driver that floats from one side of the road to the other.

Motor City . . . Detroit, Michigan.

Motor On . . . Get to movin'.

Motor Mouth . . . CB'er who is constantly talking.

Mud . . . Coffee.

Music City . . . Nashville, Tennessee.

Nap Trap . . . Motel.

Negatory . . . Negative. No.

Nickel Channel . . . On Channel 5.

O-City . . . Oklahoma City, Oklahoma.

Omni . . . Omnidirectional antenna.

Old Muddy . . . Mississippi River.

On the Side . . . Sign off. Standing by. Monitoring.

O-Town . . . Omaha, Nebraska.

Outdoors Indoors . . . Shopping mall.

Outrigger . . . Two antennas.

Over Modulate . . . Signal is too strong. Possibly illegal.

Over Your Shoulder . . . Behind you (What's it look like "over your shoulder"?).

PA . . . Public address system. External speaker.

Pair of Sevens . . . Negative contact.

Papa Bear . . . Patrolman who has "ears."

Part your Hair . . . Proceed, but at legal speed.

Pedal a Little Slower . . . Back off the accelerator pedal.
Pedal to the Metal . . . Traveling at top speed.
Pepper . . . Rocks, pebbles.
Petro Farm . . . Refinery or group of tanks.
Pickum-up . . . Pickup truck.
Piggy Bank . . . A toll station.
Pike . . . Turnpike or limited access highway.
Pit Stop . . . Gasoline stop.
Plain Wrapper . . . Unmarked patrol car. (White, blue)
Portable Stockyard . . . Livestock truck.
Put My Teeth Up for the Night . . . Go off the air (10-7).
Pressure Cooker . . . Diesel engine.
Radio Check . . . Request fellow CB'er to tell you how your radio is "getting out."
Rake the Leaves . . . Last CB'er in a convoy. (Front door "shakes the trees"–Back door "rakes the leaves.")
Ratchet Jaw . . . Very talkative CB'er.
REACT . . . Radio Emergency Associated Citizens Teams. Monitors Channel 9.
Red Wagon . . . Fire Engine.
Rest 'em Up . . . Roadside park. Rest area.
Rig . . . CB radio.
Ringing for You . . . Calling for a fellow CB'er. (Hey, "Lone Ranger," someone is "ringing for you.").
River City . . . Memphis, Tennessee.
Road Tar . . . Coffee.

Rocking Chair . . . Middle vehicle in a convoy.
Roller Skate . . . Compact car.
Rolling Revival . . . Bus or motor home type vehicle that holds chapel services at truck stops.
Rubber City . . . Akron, Ohio.
Safe Trucking . . . Have a good safe trip.
School-20 . . . School zone.
Sailboat Fuel . . . Driving on an empty tank.
See You on the Flip-flop . . . I'll talk with you when I return from my trip.
Skating Rink . . . Slick highway.
Skip . . . Talking further than 150 miles.
Smokey . . . Any policeman or patrolman.
Squawk Box . . . CB radio.
Sweet Thing . . . Lady modulator.
Short, Short . . . Quick conversation. Short span of time.
Threes (3's), 8's . . . May your day be a good one.
Two-wheeler . . . Motorcycle or bicycle.
Twister . . . Clover leaf. Highway interchange.
Uncle Charlie . . . Federal Communications Commission (FCC).
Walking All Over You . . . Your signal is being covered up by another transmission.
Walking the Dog . . . Modulating over a long distance. Clear signal.
Weight Watcher . . . Mobile inspector for Interstate Commerce Commission (ICC).
Willy Weaver . . . Intoxicated driver.
Wall to Wall . . . Signal is strong and clear.
White Knight . . . Patrolman in Georgia.

10-SIGNALS

Two sets of 10-signals are in existence, the longer, traditional set of codes and the newer set adapted by the Associated Public Safety Communications Officers, Inc. (APCO). Because conflicting information is represented by certain of the 10 codes within the two listings, both are presented here. Though it is preferred that you use the APCO code, in time of trouble it may be necessary to state the appropriate signal (10-33: APCO; 10-34: traditional) and reinforce it by stating the meaning you want to convey.

Traditional CB 10-Code

10-1 Receiving poorly
10-2 Receiving well
10-3 Stop transmitting
10-4 OK, Message received
10-5 Relay message
10-6 Busy, stand by
10-7 Out of service, leaving air
10-8 In service, subject to call
10-9 Repeat message
10-10 Transmission complete, standing by
10-11 Talking too rapidly
10-12 Visitors present
10-13 Advise weather and road condition
10-16 Make pickup at ______.
10-17 Urgent business
10-18 Anything for us?

10-19 Nothing for you, return to base
10-20 My location is _____.
10-21 Call by telephone.
10-22 Report in person to _____.
10-23 Stand by.
10-24 Completed last assignment
10-25 Can you contact?
10-26 Disregard last message.
10-27 I am moving to Channel _____.
10-28 Identify your station.
10-29 Time is up for contact.
10-30 Does not conform to FCC rules
10-32 I will give you a radio check.
10-33 Emergency traffic at this station
10-34 Trouble at this station, need help
10-35 Confidential information
10-36 Correct time is _____.
10-37 Wrecker needed at _____
10-38 Ambulance needed at _____
10-39 Your message delivered
10-41 Please tune to channel _____.
10-42 Traffic accident at _____
10-43 Traffic tieup at _____.
10-44 I have a message for you.
10-45 All units within range, please report.
10-46 Assist motorist
10-50 Break channel
10-60 What is next message number?
10-62 Unable to copy; use phone
10-63 Network directed to _____
10-64 Network clear
10-65 Awaiting your next message/assignment

10-67 All units comply.
10-69 Message received
10-70 Fire at ______
10-71 Proceed with transmission in sequence.
10-73 Speed trap at ______
10-74 Negative
10-75 You are causing interference.
10-77 Negative contact
10-81 Reserve hotel room for ______.
10-82 Reserve room for ______.
10-84 My telephone number is ______.
10-85 My address is ______.
10-89 Radio repairman needed at ______
10-90 I have TVI.
10-91 Talk closer to the mike.
10-92 Your transmission is out of adjustment.
10-93 Check my frequency on this channel.
10-94 Please give me a long count.
10-95 Transmit dead carrier for five seconds.
10-97 Check test signal.
10-99 Mission completed, all units secure
10-100 Restroom stop
10-200 Police needed at ______

New Official APCO 10-Signals Code

10-1 Signal Weak
10-2 Signal good
10-3 Stop transmitting.
10-4 Affirmative (OK)
10-5 Relay (to)

10-6 Busy
10-7 Out of service
10-8 In service
10-9 Say again
10-10 Negative
10-11 ______ on duty
10-12 Stand by (stop).
10-13 Existing conditions
10-14 Message/information
10-15 Message delivered
10-16 Reply to message
10-17 Enroute
10-18 Urgent
10-19 (in) contact
10-20 Location
10-21 Call ______ by phone
10-22 Disregard
10-23 Arrived at scene
10-24 Assignment completed
10-25 Report to (meet)
10-26 Estimated arrival time
10-27 License/permit information
10-28 Ownership information
10-29 Records check
10-30 Danger/caution
10-31 Pick up
10-32 ______ units needed–specify/number/type
10-33 Help me quick!
10-34 Time

PART FIVE

"10-99" (Mission Concluded)

It has been the intention of this book to present ideas and information which may be of value to the Christian who would use Citizens Band radio. The apostle Paul said, "I have become all things to all men that I may win some." God is not limited in the ways he uses men; we should not be hesitant to employ whatever means he provides for us to do his work.

We have discussed how individuals can use CB personally as a tool for ministering, caring, and witnessing. Perhaps it is only fitting that we offer each of you the opportunity to become involved in an international group whose purpose it is to exemplify Christian ideals through action. Only recently such an organization has been formed. It offers Christians throughout the world the opportunity to participate in the CB Ministry.

The name of this organization is "CB'ers for Christ International." It was conceived through the experience of Jerry Mitchell, pastor of the Emmanuel Baptist Church, Mansfield, Louisiana.

Jerry related that "everything that I am and have is dedicated to Christ. However, I had not committed by CB radio to Christ. When I decided to make this commitment, God gave me the inspiration for beginning CB'ers for Christ International (CBCI)."

Not yet out of its infancy, the organization is composed of only one club with 107 members and 33 CB radios. CBCI could be on its way to true national and perhaps international status with some inquiries coming from missionaries in foreign countries who have CB's. Individuals from other cities and states have also asked for information and the likelihood of associate clubs springing up is excellent.

Christian work done by CBCI has been varied and meaningful, and Mitchell pointed out some examples.

A recent tornado ripped through the Mansfield area, driving people from their homes and devastating the area. In addition, the stricken area, now easy prey for scavengers and looters, was without electricity. CBCI went into action. Within thirty minutes they had set up a base from which to work.

Generators were brought in to light up the devastated area as a protection against thievery; food and clothing were provided, and an offering was received to help those in need. Mitchell noted that "we almost ended up with more food and clothing than we could handle." Actually CBCI mobile and base units were the only help the police had in watching over the tornado-stricken area.

CBCI also will be a right arm to Toledo Bend chaplain, Jimmy Brossette, according to Mitchell. The chaplain, himself a CBCI member, and the club are laying future plans for cooperative action.

In the event of drownings and mishaps such as persons being lost, Brossette will contact CBCI, which in turn will lend its expertise to solving problems that arise.

Mitchell says that CBCI also assists stranded motorists and cites a recent case in the Mansfield area in which a lady was marooned with a broken-down automobile. CBCI came to her rescue and provided protection and companionship. The club even secured parts for the disabled vehicle.

This past July 4 weekend also saw CBCI action. A coffee break for motorists was set up at Loggy Bayou on U.S. Highway 71 south of Shreveport on July 2-3. Members set up booths and provided coffee, cookies, and Christian litarature free to motorists who stopped.

More than 250 vehicles stopped for coffee breaks. Christian literature including the New Testament, Gospel of John, and a tract, "What's Your 10-20?" (mentioned earlier) was distributed. Cars came from many different states, the most distant being a vehicle from New York.

Mitchell says the purpose of the coffee break was to "wake travelers with coffee," helping them to avoid accidents, while at the same time offering food for the soul through Christian literature. Thus, he added, we can help persons "body and soul."

Too, Mitchell and other members have used CBCI as a means of witnessing to others. Some people have actually been won to Christ through CB conversations, and Mitchell claims facts to back this up.

The club also is starting a daily devotional time. Every morning on Channel 13 "we will have a short Scripture reading and prayer to start the day with," Mitchell said. In an effort to increase services to others in the name of Christ, CBCI will monitor Channel 12 every hour of every day when the club reaches a certain growth level.

How does one join CBCI? How do interested parties get together and organize an associate club? Naturally, an applicant must first be a Christian. Then he must have a desire to use his CB to help others. If these two qualifications are already met, a prospective member may contact CB'ers for Christ, 1314 Polk Street, Mansfield, Louisiana 71052, and inquire further. Guidelines and purpose are found in the appendix.

CBCI wants to develop associate clubs in major population centers, Mitchell says. Members can be of any Christian faith–CBCI currently has members from Southern Baptist, United Pentecostal, United Methodists, Catholic, and Congregational Methodist churches–and do not have to own CB equipment. They must, however, have a deep interest in it.

Soon CBCI will be incorporated as a non-profit religious organization. When that occurs plans will be developed to purchase an emergency van and offer an emergency medical training course to members so they can be of greater service. The van and trained club members will assist in emergencies wherever they arise. When that comes we will truly be "on the move for God."

Along with Mitchell and others, we feel that reliance upon the Holy Spirit should be the prerequisite for every Christian action. We are further convicted that it is wise to follow Jesus' example and spiritual guidelines in communicating the gospel.

The ways that God uses man "have no reference to His own advantage, but to ours only; and so far as He is concerned, has reference only to His goodness. When we take pity upon a man and care for him, it is for his advantage we do so; but somehow or other our own advantage follows by a sort of natural consequence, for God does not leave the mercy we show to him who needs it to go without reward. Now this is our highest reward, that we should fully enjoy Him, and that all who enjoy Him should enjoy one another in Him."[5]

That we should fully enjoy Him—what a wonderful prize for communicating God's word!

APPENDIX

CB'ers FOR CHRIST
Post Office Box 4117
Casper, Wyoming 82601

PURPOSE: CB'ers for Christ exists to add a Christian influence to Citizens Band radio, to encourage Christian fellowship, to be a consistent Christian witness to those with whom its members talk via CB, and to observe and promote courteous, proper radio operational procedures.

CB'ers FOR CHRIST INTERNATIONAL
(CBCI)
A Christian, Family Radio Club
Mansfield, Louisiana

PURPOSE:

1. To promote Christ's kingdom.
2. To promote fellowship among God's people.
3. To assist, in Christ's name, those in distress and the church.
4. To give a witness on the airwaves for Christ and the church.
5. To give a broader scope to the ministry outside the church.
6. To encourage God's people to support the church and all its programs.

BIBLIOGRAPHY

1. *All About CB Two-Way Radio* by Radio Shack's Editorial and Technical staff, (Ft. Worth: Printed in U.S., 1976)
2. *Big Dummy's Guide to C.B. Radio* (Book Publishing Co: Summertown, Tenn., 1976) p. 52
3. Conner, W. T., *Christian Doctrine* (Nashville, Tenn.: Broadman Press, 1937) p. 14
4. *Dictionary of CB Lingo* by the Editors of Elementary Electronics, (New York: Davis Publications, Inc., 1976
5. *Great Books of the Western World, Encyclopaedia Britannica, Inc.* (London, Chicago, Toronto: 1952) Vol. 18, pp. 633, 696
6. Green, John, "CB'ers for Christ International," (unpublished article), August, 1976. Used by permission.
7. Hanlin, Don, *CB Today* (Indianapolis: Glass Publishing Corporation, October, 1976)
8. *International Standard Bible Encyclopedia,* p. 597
9. *Interpreter's Dictionary of the Bible* (New York: Abingdon Press, 1962), Vol. 4, p. 473
10. Knight, Walter B., *Knight's Master Book of New Illustrations* (Grand Rapids, Michigan: Wm. B. Eerdmans Publishing Co., 1956) p. 460
11. Miller, George A., *Language and Communication* (New York, Toronto, London: McGraw-Hill, 1951)
12. Scharlemann, Martin H., *Proclaiming the Par-*

ables (Concordia Publishing House: St. Louis, Missouri, 1963) p. 14

13. Shoemaker, Samuel M., *Beginning Your Ministry* (Harper & Row: New York, Evanston, and London, 1963) p. 53
14. Strange, E. Z., *Mississippi Press Register,* (August 11, 1976) p. 6-B
15. Ten Boom, Corrie, *The Hiding Place* (Fleming H. Revell Publishing Co.: Old Tappan. New Jersey, 1971)
16. Toffler, Alvin, *Future Shock* (Bantam Books, Inc., 1971) pp. 166, 453
17. "Town Talk, Alexandria-Pineville, Louisiana" (McCormick and Co., Inc., August 11, 1976)
18. Trueblood, Elton & Haney, David, "Target: Lifestyle Evangelism and Lifestyle Ministry" (Brotherhood Commission, Memphis, Tennessee) Adapted from Dr. Haney's article in the 1974-75 *Baptist Men's Handbook* and from his *The Idea of the Laity,* (Zondervan, 1973)
19. "What Every Licensee Should Know About the Citizens Radio Service," Federal Communications Commission (SS Bulletin 1001, May, 1976)